Aircraft Engineering: Design, Structures and Systems

Aircraft Engineering: Design, Structures and Systems

Editor: Ian Booth

NY RESEARCH
P R E S S

New York

Published by NY Research Press
118-35 Queens Blvd., Suite 400,
Forest Hills, NY 11375, USA
www.nyresearchpress.com

Aircraft Engineering: Design, Structures and Systems
Edited by Ian Booth

International Standard Book Number: 978-1-63238-589-5 (Hardback)

Cataloging-in-Publication Data

Aircraft engineering : design, structures and systems / edited by Ian Booth.
   p. cm.
Includes bibliographical references and index.
ISBN 978-1-63238-589-5
1. Aeronautics. 2. Airplanes. 3. Airplanes--Design and construction. I. Booth, Ian.
TL545 .A37 2018
629.13--dc23

Contents

Preface

The field of engineering which is concerned with the design, study, invention, maintenance, building of aircrafts and spacecrafts is known as aircraft engineering. It is further divided into two major branches known as aeronautical engineering and astronautical engineering. This field uses elements from various subjects namely control engineering, fluid mechanics, mathematics, solid mechanics, aeroelasticity, flight test, etc. This book provides comprehensive insights into the field of aircraft engineering. It is compiled in such a manner, that it will provide in-depth knowledge about the theory and practice of aircraft engineering. This book is meant for students who are looking for an elaborate reference text in this field.

A short introduction to every chapter is written below to provide an overview of the content of the book:

Chapter 1 - Aircraft is a machine capable of flying because of the force of gravity by applying dynamic lift or static lift. The size and speed of an aircraft varies with the purpose of the aircraft. The chapter on aircraft and structural dynamics offers an insightful focus, keeping in mind the complex subject matter; **Chapter 2 -** This chapter studies the various motions studied in aircraft structural dynamics. Some of the topics discussed are support motion, rotating unbalance, harmonic motion, etc. The section strategically encompasses and incorporates the major components and key concepts of the subject matter, providing a complete understanding; **Chapter 3 -** Energy damping is a process which occurs in oscillatory systems. The purpose of this process is to discharge energy from the system. The topics discussed in the chapter are of great importance to broaden the existing knowledge on aircraft engineering; **Chapter 4 -** The differential equation of motion of a dynamic system may be linear or non-linear. These vary as systems with different degrees of freedom (DOF) are taken into consideration. Aircraft engineering is best understood in confluence with the major topics listed in the following chapter.

Finally, I would like to thank my fellow scholars who gave constructive feedback and my family members who supported me at every step.

Editor

An Overview of Aircraft and Structural Dynamics

Aircraft is a machine capable of flying because of the force of gravity by applying dynamic lift or static lift. The size and speed of an aircraft varies with the purpose of the aircraft. The chapter on aircraft and structural dynamics offers an insightful focus, keeping in mind the complex subject matter.

Aircraft

An aircraft is a machine that is able to fly by gaining support from the air. It counters the force of gravity by using either static lift or by using the dynamic lift of an airfoil, or in a few cases the downward thrust from jet engines.

NASA test aircraft

The human activity that surrounds aircraft is called *aviation*. Crewed aircraft are flown by an onboard pilot, but unmanned aerial vehicles may be remotely controlled or self-controlled by onboard computers. Aircraft may be classified by different criteria, such as lift type, aircraft propulsion, usage and others.

The Mil Mi-8 is the most-produced helicopter in history

Voodoo, a modified P 51 Mustang is the 2014 Reno Air Race Champion

History

Flying model craft and stories of manned flight go back many centuries, however the first manned ascent – and safe descent – in modern times took place by larger hot-air balloons developed in the 18th century. Each of the two World Wars led to great technical advances. Consequently, the history of aircraft can be divided into five eras:

- Pioneers of flight, from the earliest experiments to 1914.

- First World War, 1914 to 1918.

- Aviation between the World Wars, 1918 to 1939.

- Second World War, 1939 to 1945.

- Postwar era, also called the jet age, 1945 to the present day.

Methods of Lift

Lighter Than Air – Aerostats

Aerostats use buoyancy to float in the air in much the same way that ships float on the water. They are characterized by one or more large gasbags or canopies, filled with a relatively low-density gas such as helium, hydrogen, or hot air, which is less dense than the surrounding air. When the weight of this is added to the weight of the aircraft structure, it adds up to the same weight as the air that the craft displaces.

Hot air balloons

Small hot-air balloons called sky lanterns were first invented in ancient China prior to the 3rd century BC and used primarily in cultural celebrations, and were only the second type of aircraft to fly, the first being kites which were first invented in ancient China over two thousand years ago.

Airship USS Akron over Manhattan in the 1930s

A balloon was originally any aerostat, while the term airship was used for large, powered aircraft designs – usually fixed-wing. In 1919 Frederick Handley Page was reported as referring to "ships of the air," with smaller passenger types as "Air yachts." In the 1930s, large intercontinental flying boats were also sometimes referred to as "ships of the air" or "flying-ships" – though none had yet been built. The advent of powered balloons, called dirigible balloons, and later of rigid hulls allowing a great increase in size, began to change the way these words were used. Huge powered aerostats, characterized by a rigid outer framework and separate aerodynamic skin surrounding the gas bags, were produced, the Zeppelins being the largest and most famous. There were still no fixed-wing aircraft or non-rigid balloons large enough to be called airships, so "airship" came to be synonymous with these aircraft. Then several accidents, such as the Hindenburg disaster in 1937, led to the demise of these airships. Nowadays a "balloon" is an unpowered aerostat and an "airship" is a powered one.

A powered, steerable aerostat is called a *dirigible*. Sometimes this term is applied only to non-rigid balloons, and sometimes *dirigible balloon* is regarded as the definition of an airship (which may then be rigid or non-rigid). Non-rigid dirigibles are characterized by a moderately aerodynamic gasbag with stabilizing fins at the back. These soon became known as *blimps*. During the Second World War, this shape was widely adopted for tethered balloons; in windy weather, this both reduces the strain on the tether and stabilizes the balloon. The nickname *blimp* was adopted along with the shape. In modern times, any small dirigible or airship is called a blimp, though a blimp may be unpowered as well as powered.

Heavier-than-air – Aerodynes

Heavier-than-air aircraft, such as airplanes, must find some way to push air or gas downwards, so that a reaction occurs (by Newton's laws of motion) to push the aircraft upwards. This dynamic movement through the air is the origin of the term *aerodyne*. There are two ways to produce dynamic upthrust: aerodynamic lift, and powered lift in the form of engine thrust.

Aerodynamic lift involving wings is the most common, with fixed-wing aircraft being kept in the air by the forward movement of wings, and rotorcraft by spinning wing-shaped rotors sometimes called rotary wings. A wing is a flat, horizontal surface, usually shaped in cross-section as an aerofoil. To fly, air must flow over the wing and generate lift. A *flexible wing* is a wing made of fabric or thin sheet material, often stretched over a rigid frame. A *kite* is tethered to the ground and relies on the speed of the wind over its wings, which may be flexible or rigid, fixed, or rotary.

With powered lift, the aircraft directs its engine thrust vertically downward. V/STOL aircraft, such as the Harrier Jump Jet and F-35B take off and land vertically using powered lift and transfer to aerodynamic lift in steady flight.

A pure rocket is not usually regarded as an aerodyne, because it does not depend on the air for its lift (and can even fly into space); however, many aerodynamic lift vehicles have been powered or assisted by rocket motors. Rocket-powered missiles that obtain aerodynamic lift at very high speed due to airflow over their bodies are a marginal case.

Fixed-wing

The forerunner of the fixed-wing aircraft is the kite. Whereas a fixed-wing aircraft relies on its forward speed to create airflow over the wings, a kite is tethered to the ground and relies on the wind blowing over its wings to provide lift. Kites were the first kind of aircraft to fly, and were invented in China around 500 BC. Much aerodynamic research was done with kites before test aircraft, wind tunnels, and computer modelling programs became available.

An Airbus A380, the world's largest passenger airliner

The first heavier-than-air craft capable of controlled free-flight were gliders. A glider designed by Cayley carried out the first true manned, controlled flight in 1853.

Practical, powered, fixed-wing aircraft (the aeroplane or airplane) were invented by Wilbur and Orville Wright. Besides the method of propulsion, fixed-wing aircraft are in general characterized by their wing configuration. The most important wing characteristics are:

- Number of wings – Monoplane, biplane, etc.

- Wing support – Braced or cantilever, rigid, or flexible.

- Wing planform – including aspect ratio, angle of sweep, and any variations along the span (including the important class of delta wings).

- Location of the horizontal stabilizer, if any.

- Dihedral angle – positive, zero, or negative (anhedral).

A variable geometry aircraft can change its wing configuration during flight.

A *flying wing* has no fuselage, though it may have small blisters or pods. The opposite of this is a *lifting body*, which has no wings, though it may have small stabilizing and control surfaces.

Wing-in-ground-effect vehicles are not considered aircraft. They "fly" efficiently close to the surface of the ground or water, like conventional aircraft during takeoff. An example is the Russian ekranoplan (nicknamed the "Caspian Sea Monster"). Man-powered aircraft also rely on ground effect to remain airborne with a minimal pilot power, but this is only because they are so underpowered—in fact, the airframe is capable of flying higher.

Rotorcraft

Rotorcraft, or rotary-wing aircraft, use a spinning rotor with aerofoil section blades (a *rotary wing*) to provide lift. Types include helicopters, autogyros, and various hybrids such as gyrodynes and compound rotorcraft.

An autogyro

Helicopters have a rotor turned by an engine-driven shaft. The rotor pushes air downward to create lift. By tilting the rotor forward, the downward flow is tilted backward, producing thrust for forward flight. Some helicopters have more than one rotor and a few have rotors turned by gas jets at the tips.

Autogyros have unpowered rotors, with a separate power plant to provide thrust. The rotor is tilted backward. As the autogyro moves forward, air blows upward across the rotor, making it spin. This spinning increases the speed of airflow over the rotor, to provide lift. Rotor kites are unpowered autogyros, which are towed to give them forward speed or tethered to a static anchor in high-wind for kited flight.

Cyclogyros rotate their wings about a horizontal axis.

Compound rotorcraft have wings that provide some or all of the lift in forward flight. They are nowadays classified as *powered lift* types and not as rotorcraft. *Tiltrotor* aircraft (such as the V-22 Osprey), tiltwing, tailsitter, and coleopter aircraft have their rotors/propellers horizontal for vertical flight and vertical for forward flight.

Other Methods of Lift

- A *lifting body* is an aircraft body shaped to produce lift. If there are any wings, they are too small to provide significant lift and are used only for stability and control. Lifting bodies are not efficient: they suffer from high drag, and must also travel at high speed to generate enough lift to fly. Many of the research prototypes, such as the Martin-Marietta X-24, which led up to the Space Shuttle, were lifting bodies (though the shuttle itself is not), and some supersonic missiles obtain lift from the airflow over a tubular body.

- *Powered lift* types rely on engine-derived lift for vertical takeoff and landing (VTOL). Most types transition to fixed-wing lift for horizontal flight. Classes of powered lift types include VTOL jet aircraft (such as the Harrier jump-jet) and tiltrotors (such as the V-22 Osprey), among others. A few experimental designs rely entirely on engine thrust to provide lift throughout the whole flight, including personal fan-lift hover platforms and jetpacks. VTOL research designs include the flying Bedstead.

- The *Flettner airplane* uses a rotating cylinder in place of a fixed wing, obtaining lift from the magnus effect.

- The *ornithopter* obtains thrust by flapping its wings.

X-24B lifting body, specialized glider

Scale - Sizes and Speeds

Sizes

The smallest aircraft are toys, and—even smaller - nano-aircraft.

The largest aircraft by dimensions and volume (as of 2016) is the 302-foot-long (about 95 meters) British Airlander 10, a hybrid blimp, with helicopter and fixed-wing features, and reportedly capable of speeds up to 90 mph (about 150 km/h), and an airborne endurance of two weeks with a payload of up to 22,050 pounds (11 tons).

The largest aircraft by weight and largest regular fixed-wing aircraft ever built (as of 2016), is the Antonov An-225. That Ukrainian-built 6-engine Russian transport of the 1980s is 84 metres (276 feet) long, with an 88-meter (289 foot) wingspan. It holds the world payload record, after transporting 428,834 pounds (200 tons) of goods, and has recently flown 100-ton loads commercially. Weighing in at somewhere between 1.1 and 1.4 million pounds (550-700 tons) maximum loaded weight, it is also the heaviest aircraft to be built, to date. It can cruise at 500 mph.

The largest military airplanes are the Ukrainian/Russian Antonov An-124 (world's second-largest airplane, also used as a civilian transport), and American Lockheed C-5

Galaxy transport, weighing, loaded, over 765,000 pounds (over 380 tons). The 8-engine, piston/propeller Hughes HK-1 "Spruce Goose," an American World War II wooden flying boat transport—with a greater wingspan (94 meters / 260 feet) than any current aircraft, and a tail-height equal to the tallest (Airbus A380-800 at 24.1 meters / 78 feet) -- flew only one short hop in the late 1940s, and never flew out of ground effect.

The largest civilian airplanes, apart from the above-noted An-225 and An-124, are the French Airbus Beluga cargo transport derivative of the Airbus A300 jet airliner, the American Boeing Dreamlifter cargo transport derivative of the Boeing 747 jet airliner/transport (the 747-200B was, at its creation in the 1960s, the heaviest aircraft ever built, with a maximum weight of 836,000 pounds (over 400 tons)), and the double-decker French Airbus A380 "super-jumbo" jet airliner (the world's largest passenger airliner).

Speeds

The fastest recorded powered aircraft flight and fastest recorded aircraft flight of an air-breathing powered aircraft was of the NASA X-43A Pegasus, a scramjet-powered, hypersonic, lifting body experimental research aircraft, at Mach 9.6 (nearly 7,000 mph). The X-43A set that new mark, and broke its own world record (of Mach 6.3, nearly 5,000 mph, set in March, 2004) on its third and final flight on Nov. 16, 2004.

Prior to the X-43A, the fastest recorded powered airplane flight (and still the record for the fastest manned, powered airplane / fastest manned, non-spacecraft aircraft) was of the North American X-15A-2, rocket-powered airplane at 4,520 mph (7,274 km/h), Mach 6.72, on October 3, 1967. On one flight it reached an altitude of 354,300 feet.

The fastest known, production aircraft (other than rockets and missiles) currently or formerly operational (as of 2016) are:

- The fastest fixed-wing aircraft, and fastest glider, is the Space Shuttle, a rocket-glider hybrid, which has re-entered the atmosphere as a fixed-wing glider at over Mach 25 (over 25 times the speed of sound—about 17,000 mph at re-entry to Earth's atmosphere).

- The fastest military airplane ever built: Lockheed SR-71 Blackbird, a U.S. reconnaissance jet fixed-wing aircraft, known to fly beyond Mach 3.3 (about 2,200 mph at cruising altitude). On July 28, 1976, an SR-71 set the record for the fastest and highest-flying operational aircraft with an absolute speed record of 2,193 mph and an absolute altitude record of 85,068 feet. At its retirement in the January 1990, it was the fastest air-breathing aircraft / fastest jet aircraft in the world—a record still standing as of August, 2016.

  Note: Some sources refer to the above-mentioned X-15 as the "fastest military airplane" because it was partly a project of the U.S. Navy and Air Force; however, the X-15 was not used in non-experimental actual military operations.

- The fastest current military aircraft are the Soviet/Russian MiG-25—capable of Mach 3.2 (2,170 mph), at the expense of engine damage, or Mach 2.83 (1,920 mph) normally—and the Russian MiG-31E (also capable of Mach 2.83 normally). Both are fighter-interceptor jet airplanes, in active operations as of 2016.

- The fastest civilian airplane ever built, and fastest passenger airliner ever built: the briefly operated Tupolev Tu-144 supersonic jet airliner (Mach 2.35, 1,600 mph, 2,587 km/h), which was believed to cruise at about Mach 2.2. The Tu-144 (officially operated from 1968 to 1978, ending after two crashes of the small fleet) was outlived by its rival, the Concorde SST (Mach 2.23), a French/British supersonic airliner, known to cruise at Mach 2.02 (1.450 mph, 2,333 kmh at cruising altitude), operating from 1976 until the small Concorde fleet was grounded permanently in 2003, following the crash of one in the early 2000s.

- The fastest civilian airplane currently flying: the Cessna Citation Ten, an American business jet, capable of Mach 0.935 (over 600 mph at cruising altitude). Its rival, the American Gulfstream 650 business jet, can reach Mach 0.925.

- The fastest airliner currently flying is the Boeing 747, quoted as being capable of cruising over Mach 0.885 (over 550 mph). Previously, the fastest were the troubled, short-lived Russian (Soviet Union) Tupolev Tu-144 SST (Mach 2.35) and the French/British Concorde SST (Mach 2.23, normally cruising at Mach 2) . Before them, the Convair 990 Coronado jet airliner of the 1960s flew at over 600 mph.

Propulsion

Unpowered Aircraft

Gliders are heavier-than-air aircraft that do not employ propulsion once airborne. Take-off may be by launching forward and downward from a high location, or by pulling into the air on a tow-line, either by a ground-based winch or vehicle, or by a powered "tug" aircraft. For a glider to maintain its forward air speed and lift, it must descend in relation to the air (but not necessarily in relation to the ground). Many gliders can 'soar' – gain height from updrafts such as thermal currents. The first practical, controllable example was designed and built by the British scientist and pioneer George Cayley, whom many recognise as the first aeronautical engineer. Common examples of gliders are sailplanes, hang gliders and paragliders.

Balloons drift with the wind, though normally the pilot can control the altitude, either by heating the air or by releasing ballast, giving some directional control (since the wind direction changes with altitude). A wing-shaped hybrid balloon can glide directionally when rising or falling; but a spherically shaped balloon does not have such directional control.

Kites are aircraft that are tethered to the ground or other object (fixed or mobile) that maintains tension in the tether or kite line; they rely on virtual or real wind blowing over and under them to generate lift and drag. Kytoons are balloon-kite hybrids that are shaped and tethered to obtain kiting deflections, and can be lighter-than-air, neutrally buoyant, or heavier-than-air.

Powered Aircraft

Powered aircraft have one or more onboard sources of mechanical power, typically aircraft engines although rubber and manpower have also been used. Most aircraft engines are either lightweight piston engines or gas turbines. Engine fuel is stored in tanks, usually in the wings but larger aircraft also have additional fuel tanks in the fuselage.

Propeller Aircraft

A turboprop-engined DeHavilland Twin Otter adapted as a floatplane

Propeller aircraft use one or more propellers (airscrews) to create thrust in a forward direction. The propeller is usually mounted in front of the power source in *tractor configuration* but can be mounted behind in *pusher configuration*. Variations of propeller layout include *contra-rotating propellers* and *ducted fans*.

Many kinds of power plant have been used to drive propellers. Early airships used man power or steam engines. The more practical internal combustion piston engine was used for virtually all fixed-wing aircraft until World War II and is still used in many smaller aircraft. Some types use turbine engines to drive a propeller in the form of a turboprop or propfan. Human-powered flight has been achieved, but has not become a practical means of transport. Unmanned aircraft and models have also used power sources such as electric motors and rubber bands.

Jet Aircraft

Jet aircraft use airbreathing jet engines, which take in air, burn fuel with it in a combustion chamber, and accelerate the exhaust rearwards to provide thrust.

Lockheed Martin F-22A Raptor

Turbojet and turbofan engines use a spinning turbine to drive one or more fans, which provide additional thrust. An afterburner may be used to inject extra fuel into the hot exhaust, especially on military "fast jets". Use of a turbine is not absolutely necessary: other designs include the pulse jet and ramjet. These mechanically simple designs cannot work when stationary, so the aircraft must be launched to flying speed by some other method. Other variants have also been used, including the motorjet and hybrids such as the Pratt & Whitney J58, which can convert between turbojet and ramjet operation.

Compared to propellers, jet engines can provide much higher thrust, higher speeds and, above about 40,000 ft (12,000 m), greater efficiency. They are also much more fuel-efficient than rockets. As a consequence nearly all large, high-speed or high-altitude aircraft use jet engines.

Rotorcraft

Some rotorcraft, such as helicopters, have a powered rotary wing or *rotor*, where the rotor disc can be angled slightly forward so that a proportion of its lift is directed forwards. The rotor may, like a propeller, be powered by a variety of methods such as a piston engine or turbine. Experiments have also used jet nozzles at the rotor blade tips.

Other Types of Powered Aircraft

- *Rocket-powered aircraft* have occasionally been experimented with, and the Messerschmitt *Komet* fighter even saw action in the Second World War. Since then, they have been restricted to research aircraft, such as the North American X-15, which traveled up into space where air-breathing engines cannot work (rockets carry their own oxidant). Rockets have more often been used as a supplement to the main power plant, typically for the rocket-assisted take off of heavily loaded aircraft, but also to provide high-speed dash capability in some hybrid designs such as the Saunders-Roe SR.53.

- The *ornithopter* obtains thrust by flapping its wings. It has found practical use in a model hawk used to freeze prey animals into stillness so that they can be captured, and in toy birds.

Design and construction

Aircraft are designed according to many factors such as customer and manufacturer demand, safety protocols and physical and economic constraints. For many types of aircraft the design process is regulated by national airworthiness authorities.

The key parts of an aircraft are generally divided into three categories:

- The *structure* comprises the main load-bearing elements and associated equipment.

- The *propulsion system* (if it is powered) comprises the power source and associated equipment, as described above.

- The *avionics* comprise the control, navigation and communication systems, usually electrical in nature.

Structure

The approach to structural design varies widely between different types of aircraft. Some, such as paragliders, comprise only flexible materials that act in tension and rely on aerodynamic pressure to hold their shape. A balloon similarly relies on internal gas pressure but may have a rigid basket or gondola slung below it to carry its payload. Early aircraft, including airships, often employed flexible doped aircraft fabric covering to give a reasonably smooth aeroshell stretched over a rigid frame. Later aircraft employed semi-monocoque techniques, where the skin of the aircraft is stiff enough to share much of the flight loads. In a true monocoque design there is no internal structure left.

The key structural parts of an aircraft depend on what type it is.

Aerostats

Lighter-than-air types are characterised by one or more gasbags, typically with a supporting structure of flexible cables or a rigid framework called its hull. Other elements such as engines or a gondola may also be attached to the supporting structure.

Aerodynes

Heavier-than-air types are characterised by one or more wings and a central fuselage. The fuselage typically also carries a tail or empennage for stability and control, and an undercarriage for takeoff and landing. Engines may be located on the fuselage or wings. On a fixed-wing aircraft the wings are rigidly attached to the fuselage, while on a rotorcraft the wings are attached to a rotating vertical shaft. Smaller designs sometimes use flexible materials for part or all of the structure, held in place either by a rigid frame or by air pressure. The fixed parts of the structure comprise the airframe.

Airframe diagram for an AgustaWestland AW101 helicopter

Avionics

The avionics comprise the flight control systems and related equipment, including the cockpit instrumentation, navigation, radar, monitoring, and communication systems.

Flight Characteristics

Flight Envelope

The flight envelope of an aircraft refers to its capabilities in terms of airspeed and load factor or altitude. The term can also refer to other measurements such as maneuverability. When a craft is pushed, for instance by diving it at high speeds, it is said to be flown *outside the envelope*, something considered unsafe.

Range

The Boeing 777-200LR is the longest-range airliner, capable of flights of more than halfway around the world.

The range is the distance an aircraft can fly between takeoff and landing, as limited by the time it can remain airborne.

For a powered aircraft the time limit is determined by the fuel load and rate of consumption.

For an unpowered aircraft, the maximum flight time is limited by factors such as weather conditions and pilot endurance. Many aircraft types are restricted to daylight hours, while balloons are limited by their supply of lifting gas. The range can be seen as the average ground speed multiplied by the maximum time in the air.

Flight Dynamics

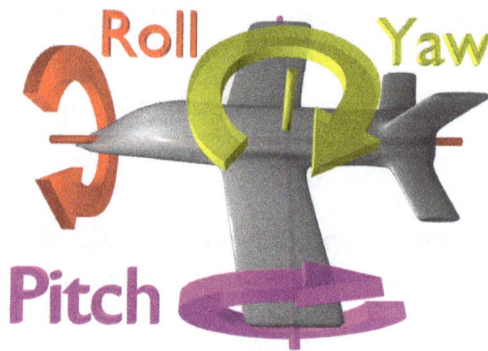

Flight dynamics is the science of air vehicle orientation and control in three dimensions. The three critical flight dynamics parameters are the angles of rotation around three axes about the vehicle's center of mass, known as *pitch, roll,* and *yaw* (quite different from their use as Tait-Bryan angles).

- Roll is a rotation about the longitudinal axis (equivalent to the rolling or heeling of a ship) giving an up-down movement of the wing tips measured by the roll or bank angle.

- Pitch is a rotation about the sideways horizontal axis giving an up-down movement of the aircraft nose measured by the angle of attack.

- Yaw is a rotation about the vertical axis giving a side-to-side movement of the nose known as sideslip.

Flight dynamics is concerned with the stability and control of an aircraft's rotation about each of these axes.

Stability

An aircraft that is unstable tends to diverge from its current flight path and so is difficult to fly. A very stable aircraft tends to stay on its current flight path and is difficult to manoeuvre—so it is important for any design to achieve the desired degree of stability. Since the widespread use of digital computers, it is increasingly common for designs to be inherently unstable and rely on computerised control systems to provide artificial stability.

The tail assembly of a Boeing 747–200

A fixed wing is typically unstable in pitch, roll, and yaw. Pitch and yaw stabilities of conventional fixed wing designs require horizontal and vertical stabilisers, which act similarly to the feathers on an arrow. These stabilizing surfaces allow equilibrium of aerodynamic forces and to stabilise the flight dynamics of pitch and yaw. They are usually mounted on the tail section (empennage), although in the canard layout, the main aft wing replaces the canard foreplane as pitch stabilizer. Tandem wing and Tailless aircraft rely on the same general rule to achieve stability, the aft surface being the stabilising one.

A rotary wing is typically unstable in yaw, requiring a vertical stabiliser.

A balloon is typically very stable in pitch and roll due to the way the payload is hung underneath.

Control

Flight control surfaces enable the pilot to control an aircraft's flight attitude and are usually part of the wing or mounted on, or integral with, the associated stabilizing surface. Their development was a critical advance in the history of aircraft, which had until that point been uncontrollable in flight.

Aerospace engineers develop control systems for a vehicle's orientation (attitude) about its center of mass. The control systems include actuators, which exert forces in various directions, and generate rotational forces or moments about the aerodynamic center of the aircraft, and thus rotate the aircraft in pitch, roll, or yaw. For example, a pitching moment is a vertical force applied at a distance forward or aft from the aerodynamic center of the aircraft, causing the aircraft to pitch up or down. Control systems are also sometimes used to increase or decrease drag, for example to slow the aircraft to a safe speed for landing.

The two main aerodynamic forces acting on any aircraft are lift supporting it in the air and drag opposing its motion. Control surfaces or other techniques may also be used to affect these forces directly, without inducing any rotation.

Impacts of Aircraft Use

Aircraft permit long distance, high speed travel and may be a more fuel efficient mode of transportation in some circumstances. Aircraft have environmental and climate impacts beyond fuel efficiency considerations, however. They are also relatively noisy compared to other forms of travel and high altitude aircraft generate contrails, which experimental evidence suggests may alter weather patterns.

Uses for Aircraft

Aircraft are produced in several different types optimized for various uses; military aircraft, which includes not just combat types but many types of supporting aircraft, and civil aircraft, which include all non-military types, experimental and model.

Military

Boeing B-17E in flight

A military aircraft is any aircraft that is operated by a legal or insurrectionary armed service of any type. Military aircraft can be either combat or non-combat:

- Combat aircraft are aircraft designed to destroy enemy equipment using its own armament. Combat aircraft divide broadly into fighters and bombers, with several in-between types such as fighter-bombers and ground-attack aircraft (including attack helicopters).

- Non-combat aircraft are not designed for combat as their primary function, but may carry weapons for self-defense. Non-combat roles include search and rescue, reconnaissance, observation, transport, training, and aerial refueling. These aircraft are often variants of civil aircraft.

Most military aircraft are powered heavier-than-air types. Other types such as gliders and balloons have also been used as military aircraft; for example, balloons were used for observation during the American Civil War and World War I, and military gliders were used during World War II to land troops.

Civil

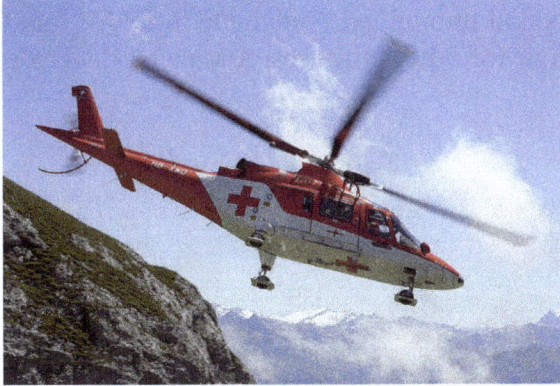

Agusta A109 helicopter of the Swiss air rescue service

Civil aircraft divide into *commercial* and *general* types, however there are some overlaps.

Commercial aircraft include types designed for scheduled and charter airline flights, carrying passengers, mail and other cargo. The larger passenger-carrying types are the airliners, the largest of which are wide-body aircraft. Some of the smaller types are also used in general aviation, and some of the larger types are used as VIP aircraft.

General aviation is a catch-all covering other kinds of private (where the pilot is not paid for time or expenses) and commercial use, and involving a wide range of aircraft types such as business jets (bizjets), trainers, homebuilt, gliders, warbirds and hot air balloons to name a few. The vast majority of aircraft today are general aviation types.

Experimental

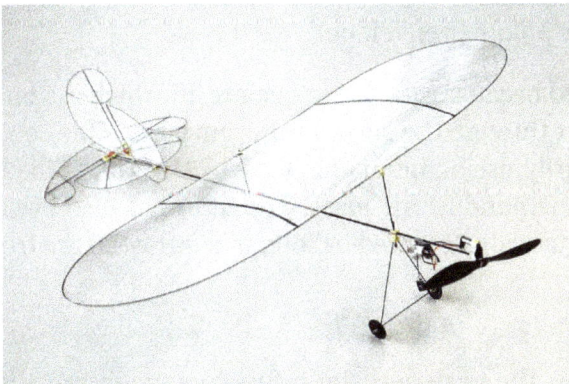

A model aircraft, weighing six grams

An experimental aircraft is one that has not been fully proven in flight, or that carries an FAA airworthiness certificate in the "Experimental" category. Often, this implies that the aircraft is testing new aerospace technologies, though the term also refers to amateur and kit-built aircraft—many based on proven designs.

Model

A model aircraft is a small unmanned type made to fly for fun, for static display, for aerodynamic research or for other purposes. A scale model is a replica of some larger design.

Aerostat

A modern aerostat used by the United States Department of Homeland Security

An aerostat is a lighter than air craft that gains its lift through the use of a buoyant gas. Aerostats include unpowered balloons and powered airships. A balloon may be free-flying or tethered. The average density of the craft is lower than the density of atmospheric air, because its main component is one or more gasbags, a lightweight skin containing a lifting gas to provide buoyancy, to which other components such as a gondola containing equipment or people are attached. Especially with airships, the gasbags are often protected by an outer envelope.

Aerostats are so named because they use aerostatic lift which is a buoyant force that does not require movement through the surrounding air mass. This contrasts with the heavy aerodynes that primarily use aerodynamic lift which requires the movement of a wing surface through the surrounding air mass. The term has also been used in a narrower sense, to refer to the statically tethered balloon in contrast to the free-flying airship.

Terminology

In conventional usage, the term aerostat refers to any aircraft that remains aloft primarily using aerostatic buoyancy.

Historically, all aerostats were called balloons. Powered types capable of horizontal flight were referred to as dirigible balloons or simply dirigibles (from the French *dirigible* meaning steerable). These powered aerostats later came to be called airships, with the term "balloon" reserved for unpowered types, whether tethered or free-floating.

More recently, the US Government Accountability Office has used the term "aerostat" in a different sense, to distinguish the statically tethered balloon from the free-flying airship.

Types

Balloons

A free-flying hot air balloon

A balloon is an aerostat which has no means of propulsion and must be either tethered on a long cable or allowed to drift freely with the wind.

Although a free balloon travels at the speed of the wind, it is travelling with the wind so to a passenger the air feels calm and windless. To change its altitude above ground it must either adjust the amount of lift or discard ballast weight. Notable uses of free-flying balloons include meteorological balloons and sport balloons.

A tethered balloon is held down by one or more mooring lines or tethers. It has sufficient lift to hold the line taut and its altitude is controlled by winching the line in or out. A tethered balloon does feel the wind. A round balloon is unstable and bobs about in strong winds, so the kite balloon was developed with an aerodynamic shape similar to a non-rigid airship. Both kite balloons and non-rigid airships are sometimes called "blimps". Notable uses of tethered balloons include observation balloons and barrage balloons.

Airships

An airship is a powered, free-flying aerostat that can be steered. Airships divide into rigid, semi-rigid and non-rigid types, with these last often known as Blimps.

A rigid airship has an outer framework or skin surrounding the lifting gas bags inside it, The outer envelope keeps its shape even if the gasbags are deflated. The great Zeppelin airships of the twentieth century were rigid types.

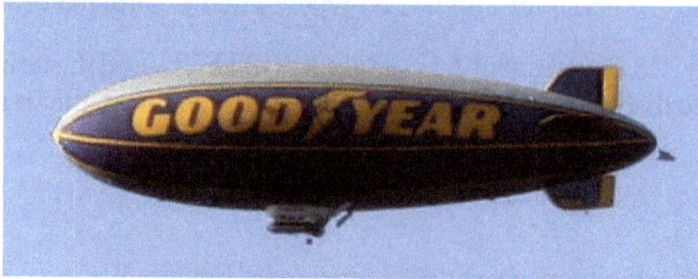

The Goodyear blimps are non-rigid airships

A non-rigid airship or blimp deflates like a balloon as it loses gas. The Goodyear blimps are still a common sight in the USA.

A semi-rigid airship has a deflatable gas bag like a non-rigid but with a supporting structure to help it hold its shape while aloft. The first practical airship, the Santos-Dumont No.6 was a semi-rigid.

Some airships obtain additional lift aerodynamically as they travel through the air, using the shape of their envelope or through the addition of fins or even small wings. Types designed to exploit this lifting effect in normal cruise are called hybrid airships.

Hybrid Aerostats

A hybrid type uses both static buoyancy and dynamic airflow to provide lift. The dynamic movement may be created either using propulsive power as a hybrid airship or by tethering in the wind like a kite, as a kytoon.

Piasecki Helicopter developed the Piasecki PA-97 Helistat using the rotor systems from four obsolete helicopters and a surplus Navy blimp, in order to provide a capability to lift heavier loads than a single helicopter could provide. The aircraft suffered a fatal accident during a test flight. In 2008, Boeing and SkyHook International resurrected the concept and announced a proposed design of the SkyHook JHL-40.

The Allsopp Helikite is a combination of a helium balloon and a kite to form a single, aerodynamically sound tethered aircraft, that exploits both wind and helium for its lift.

Lifting Gases

In order to provide buoyancy, any lifting gas must be lighter, i.e. less dense, than the surrounding air. A hot air balloon is open at the bottom to allow hot air to enter, while the gas balloon is closed to stop the (cold) lifting gas from escaping. Common lifting gases have included hydrogen, coal gas and helium.

Hot Air

When heated, air expands. This lowers its density and creates lift. Small hot air balloons or lanterns have been flown in China since ancient times. The first modern man-lifting aerostat, made by the Montgolfier brothers, was a hot air balloon. Most early balloons however were gas balloons. Interest in the sport of hot air ballooning reawoke in the second half of the twentieth century and even some hot-air airships have been flown.

Hydrogen

Hydrogen is the lightest of all gases and a manned hydrogen balloon was flown soon after the Montgolfier brothers. There is no need to burn fuel, so a gas balloon can stay aloft far longer than a hot-air balloon. It is also safer if there is no flame on board, since the materials used to make aerostats are flammable. Hydrogen soon became the most common lifting gas for both balloons and, later, airships. But hydrogen itself is flammable and, following several major disasters in the 1930s, it fell out of use.

Coal Gas

Coal gas comprises a mix of methane and other gases, and typically has about half the lifting power of hydrogen. In the late nineteenth and early twentieth centuries municipal gas works became common and provided a cheap source of lifting gas. Some works were able to produce a special mix for ballooning events, incorporating a higher proportion of hydrogen and less carbon monoxide, to improve its lifting power.

Helium

Helium is the only lifting gas which is both non-flammable and non-toxic, and it has almost as much (about 92%) lifting power as hydrogen. It was not discovered in quantity until early in the twentieth century, and for many years only the USA had enough to use in airships. Almost all gas balloons and airships now use helium.

Low Pressure Gases

Although not currently practical, it may be possible to construct a rigid, lighter-than-air structure which, rather than being inflated with air, is at a vacuum relative to the surrounding air. This would allow the object to float above the ground without any heat or special lifting gas, but the structural challenges of building a rigid vacuum chamber lighter than air are quite significant. Even so, it may be possible to improve the performance of more conventional aerostats by trading gas weight for structural weight, combining the lifting properties of the gas with vacuum and possibly heat for enhanced lift.

Vibration Theory

History of vibration starts with the invention of the musical instrument like whistle and drum. Musical instruments are appreciated by Chinese, Japanese, Hindus and Egyptians as long ago as 4000 B.C.. The philosopher and mathematician Pythagoras (582-507 B.C.) is the first person to conduct scientific experiments on vibrating strings using simple apparatus called a monochord. Though the concept of pitch of sound was developed by the time of Pythagoras the relation between the pitch and frequency was not clear until the sixteenth century. The oldest book available on the subject of music is written by Aristoxenus, a pupil of Atistotle and the name of the book is 'Elements of Harmony' (320 B.C.). After this Euclid wrote 'Introduction of Harmony' in 300 B.C. In the year 20 B.C. famous roman architect Vitruvius wrote 'D Architectur Libri Decem' on the acoustic properties of theaters.

Understanding vibration involving acoustic effects like designing a musical instrument or an auditorium is followed by the invention of the worlds first seismograph in the year 132 A.D.

Replica of Zhang Heng'sseismoscope

Seismograph is an instrument to measure earthquakes. The famous historian and astronomer, Zhang Heng, invented the wine-jar shaped instrument.

Afterwards in the seventeenth century, founder father of modern experimental science Galileo Galilei (1564-1642) discussed about vibrating bodies in his book 'Discourses Concerning Two New Sciences' published in 1638. In this book he has discussed about the relationship between frequency, length, tension and density of a vibrating stretched string. The book 'Harmonicorum Libre' (1636) authored by the father of acoustics Marin Mersenne (1588-1648), came two year before because of the prohibition imposed by the Inquisitor of Rome until 1638. This experimental study is further carried by Robert Hooke (1653-1703) to find the relation between the pitch and frequency of vibration of a string. Later Joseph Sauveur (1653-1716) coined the term "acoustics", John Wallis (1616-17030) observed the mode shape phenomenon.

The foundation stone of theoretical study of vibration is layed by Brook Taylor (1685-1731). Natural frequency of vibration derived by Taylor matches well with experimental values observed by Galileo Galilei and Mersenne. The derivation is further enhanced by Daniel Bernoulli (1700-17820), Jean D'Alembert (1717-1783) and Leonard Euler (1707-1783). Credit of studying first time the vibration of thin beams with different boundary conditions goes to the famous mathematicians Euler (1744) and Bernoulli (1751) and their modeling is known as Euler-Bernoulli thin beam theory. Charles Coulomb have studied and derived the equation of motion for the torsional vibration of the suspended cylinder first time in the year 1784. He found that the period of oscillation is independent of the angle of twist.

In case of theory related to the vibration of plate, it is first derived by the German scientist E.F.F. Chaldni (1756-1824). It is later corrected by G.R. Kirchhoff (1824-1887) by providing the proper boundary conditions in the year 1850. In relation to this the problem related to the vibrating membrane (sound emitted by drum) is first studied by Simon Poisson (1781-1840). In 1877 Lord Baron Rayleigh published his classic book on 'Theory of Sound'. The effort continues with the contributions of the famous scientists like Stodola, De Laval, Timoshenko, Mindlin and many other scientist.

From the history of development of theory of vibration we may start thinking about the inspiration behind the development. In a human body, three most prime interactions are through eye, ear and voice. Eye can differentiate between the different frequencies and wave lengths and make us see anything. Ear absorbs sounds of different frequencies and distinguish those properly from each other. In case of our voice we vibrate our voice-box in different frequencies and amplitude with proper co-ordination to communicate. If we look back to the activities we were discussing, all are time dependent phenomenon and also periodic in nature. These may be categorized as the type of dynamic behaviour where oscillation takes place about a certain equilibrium position and it may be defined as the vibratory motion or vibration. To understand the properties and characteristics related to vibration we need to model a physical system. A physical system in general is very complex in nature and difficult to analyse. It consists of various components. We need to identify the physical properties of the components. More precisely the properties governing the dynamic behaviour of the system must be modeled mathematically. Parameters are defined to represent physical properties of a system. A physical system is continuous with distributed parameters. There are two different ways of modeling a physical system a) lumped parameter system and b) continuous or distributed parameter system.

Considering a damping proportional to the velocity of the mass we get a second order equation as follows

$$m\ddot{x} + c\dot{x} + kx = 0$$

Considering a trial solution, $x = Ae^{\lambda t}$; $\dot{x} = A\lambda e^{\lambda t}$; $\ddot{x} = A\lambda^2 e^{\lambda t}$

We get, $\left(m\lambda^2 + c\lambda + k\right)Ae^{\lambda t} = 0 \Rightarrow m\lambda^2 + C\lambda + K = 0$

$$\lambda^2 + \frac{c}{m}\lambda + \frac{k}{m} = 0$$

$$\Rightarrow \lambda = -\frac{c}{2m} \pm \left[\left(\frac{c}{2m}\right)^2 - \left(\frac{k}{m}\right)\right]^{1/2}$$

$$x = e^{\left(-c/2m\right)t}\left[Ae^{\sqrt{\left(c/2m\right)^2 - \left(k/m\right)}t} + Be^{-\sqrt{\left(c/2m\right)^2 - \left(k/m\right)}t}\right]$$

There are three cases possible

Case I $\left(\frac{c}{2m}\right)^2 > \frac{k}{m}$ (Over damped condition)

Solution, $x = A_1 e^{\lambda_1 t} + A_2 e^{\lambda_2 t}$ where λ_1, λ_2 are real roots which shows that it will decay (without any oscillation).

This equation shows that there is a usual creeping back of the mass to the equilibrium-position. This is the case of over damped situation and no oscillation is possible.

Case II $\left(\frac{c}{2m}\right)^2 < \frac{k}{m}$

$\rightarrow$ under damped case

or

         FREE DAMPED VIBRATION

-   There is oscillation

-   λ_1, λ_2 are complex

$$\lambda_{1,2} = -\frac{c}{2m} \pm i\left[\frac{k}{m} - \left(\frac{c}{2m}\right)^2\right]^{\frac{1}{2}} \qquad where\ i = \sqrt{-1}$$

$$\left.\begin{array}{l} \lambda_1 = -\dfrac{c}{2m} + i\omega_d \\[2mm] \lambda_2 = -\dfrac{c}{2m} + i\omega_d \end{array}\right\}_{\omega_d} = \left[\frac{k}{m} - \left(\frac{c}{2m}\right)^2\right]^{\frac{1}{2}}.$$

Note that $e^{i\theta} = \cos\theta + i\sin\theta$

$$e^{-i\theta} = \cos\theta - i\sin\theta$$

$$x = e^{-\frac{c}{2m}t}\left[A_1' e^{i\omega_d t} + A_2' e^{-\omega_d t}\right]$$

$$= e^{-\frac{c}{2m}t}\left[A_1'\left(\cos\omega_d t + i\sin\omega_d t\right) + A_2'\left(\cos\omega_d t - \sin\omega_d t\right)\right]$$

$$= e^{-\frac{c}{2m}t}\left[\left(A_1' + A_2'\right)\cos\omega_d t + i\left(A_1' - A_2'\right)\sin\omega_d t\right]$$

Assume,
$$A_1' + A_2' = A\sin\phi$$
$$i\left(A_1' + A_2'\right) = A\cos\phi$$

$$\Rightarrow x = \exp\left(-\frac{c}{2m}t\right)\left[A\sin\phi\cos\omega_d t + A\cos\phi\sin\omega_d t\right]$$

$$= \exp\left(-\frac{c}{2m}t\right)\left[A\sin\left(\omega_d t + \phi\right)\right]$$

Where
$$A^? = \left(A_1' + A_2'\right)^2 - \left(A_1' - A_2'\right)^2$$
$$= A_1'^2 + A_2'^2 + 2A_1'A_2' - A_1'^2 - A_2'^2 + 2A_1'A_2'$$
$$= 4A_1'A_2'$$
$$A = 2\left(A_1'A_2'\right)^{\frac{1}{2}}$$

$$\tan\phi = \frac{A_1' + A_2'}{i\left(A_1' + A_2'\right)}$$

Therefore, Oscillation is present but damped due to the term $\exp\left(-\dfrac{c}{2m}t\right)$.

Case III

$$\left(\frac{c}{2m}\right)^2 = \frac{k}{m} \rightarrow \text{Case of CRITICAL DAMPING}$$

Critical damping is the value of the damping coefficient c at the change over from creeping motion case I to the underdamped vibration of case II.

Critical damping $\left(\dfrac{c_c}{2m}\right)^2 = \dfrac{k}{m} \leftarrow$ frequency term

$$c_c^2 = 4km \Rightarrow c_c = 2\sqrt{km}$$

$$\omega_n = \sqrt{\frac{k}{m}} \rightarrow c_c = 2m\,\omega_n \rightarrow c_c = \frac{2k}{\omega_n}$$

$$c = \zeta \cdot \frac{2k}{\omega_n}$$

Here ζ is considered as the damping ratio, which is also referred by ζ.

Damping ratio $\zeta = \dfrac{c}{c_c} = \dfrac{Actual\ damping\ coefficient}{Critical\ damping\ coefficient}$

$$\frac{c}{2m} = \frac{\zeta c_c}{2m} = \frac{\zeta 2m\omega_n}{2m} = \zeta\omega_n$$

$$\frac{c}{2m} = \xi\omega_n$$

$$x = Ae^{-\xi\omega_n t}\sin\left(\omega_d t + \alpha\right) \text{ where, } \omega_d = \omega_n\sqrt{1-\zeta^2}$$

$$T_1 = \frac{2\pi}{\omega_d} = \frac{2\pi}{\omega_n\sqrt{1-\zeta^2}}$$

$$\lambda_{1,2} = -\zeta\,\omega_n \pm \sqrt{\left(\zeta\,\omega_n\right)^2 - \omega_n^2}$$

$$= \left(-\zeta \pm \sqrt{\zeta^2 - 1}\right)\omega_n$$

All damped free vibration, $\zeta < 1$ Case – II

$\xi = 1$, Critical damping condition Case – III

$\xi > 1$, over damped condition Case – I

Oscillatory Motion $(\zeta < 1.0)$ Case – II

$$x = e^{-\zeta\omega_n t}\left(A e^{i\sqrt{1-\zeta^2}\omega_n t} + B e^{-i\sqrt{1-\zeta^2}\omega_n t}\right)$$

The above equation can also be written in either of the following two forms

$$x = X e^{-\zeta\omega_n t}\sin\left(\sqrt{1-\zeta^2}\,\omega_n t + \phi\right)$$

$$= e^{-\zeta\omega_n t}\left(c_1 \sin\sqrt{1-\zeta^2}\,\omega_n t + c_2 \cos\sqrt{1-\zeta^2}\,\omega_n t\right)$$

Where the arbitrary constants X, ϕ or c_1 ,c_2 are determined from initial conditions at

$$t = 0\, x(0), \dot{x}(0)$$

$$\dot{x} = -\zeta\omega_n\, e^{-\zeta\omega_n t}\left(c_1 \sin\sqrt{1-\zeta^2}\,\omega_n t + c_2 \cos\sqrt{1-\zeta^2}\,\omega_n t\right)$$

$$+ e^{-\zeta\omega_n t}\left(\sqrt{1-\zeta^2}\,\omega_n\, c_1 \cos\sqrt{1-\zeta^2}\,\omega_n t - \sqrt{1-\zeta^2}\,\omega_n\, c_2 \sin\sqrt{1-\zeta^2}\,\omega_n t\right)$$

at $t = 0$ $x(0) = c_2$

$$\dot{x}(0) = -\zeta\omega_n\,(c_2) + \sqrt{1-\zeta^2}\,\omega_n\, c_1$$

$$\Rightarrow \dot{x}(0) = -\zeta\omega_n.x(0) + \sqrt{1-\zeta^2}\,\omega_n\, c_1$$

or, $c_1 = \dfrac{\dot{x}(0) + \zeta\,\omega_n\, x(0)}{\omega_n\sqrt{1-\zeta^2}}$

$$\therefore\; x = e^{-\zeta\omega_n t}\left\{\dfrac{\dot{x}(0) + \zeta\,\omega_n\, x(0)}{\omega_n\sqrt{1-\zeta^2}}\sin\sqrt{1-\zeta^2}\,\omega_n t + x(0)\cos\sqrt{1-\zeta^2}\,\omega_n t\right\}$$

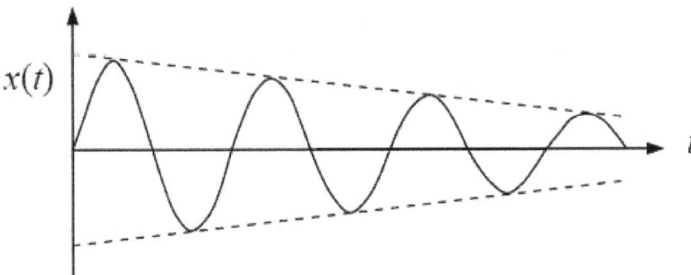

Critically damped motion $\left[\zeta = 1.0\right]$ Case III

We obtain a double root $s_1 = s_2 = -\omega_n$

The correct general solution is

$$x = \left(A + Bt\right)e^{-\omega_n t}$$

Initial conditions $x(0), \dot{x}(0)$ $\ at\ \ t = 0$

$$\dot{x} = -\omega_n A e^{-\omega_n t} + B e^{-\omega_n t} - B \omega_n t e^{-\omega_n t}$$

at $t = 0$ $\ x(0) = A$

at $t = 0$ $\ \dot{x}(0) = -\omega_n A + B$

$$\Rightarrow B = \dot{x}(0) + \omega_n A$$

$$= \dot{x}(0) + \omega_n x(0)$$

$$\therefore x = \left[x(0) + \{\dot{x}(0) + \omega_n x(0)\}t\right]e^{-\omega_n t}$$

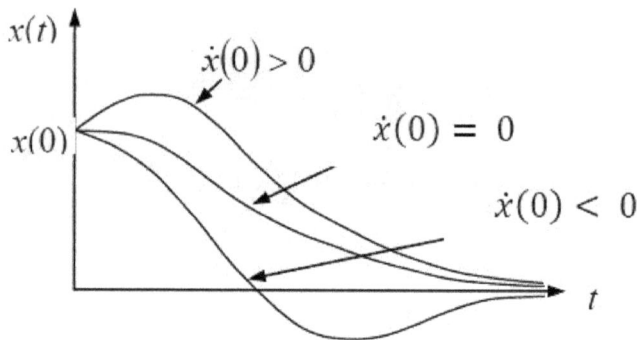

Oscillatory Motion $\left[\zeta > 1.0\right]$ Case I

$$x = Ae^{\left(-\zeta + \sqrt{\zeta^2 - 1}\right)\omega_n t} Be^{\left(-\zeta - \sqrt{\zeta^2 - 1}\right)\omega_n t}$$

$$\dot{x} = A\left(-\zeta + \sqrt{\zeta^2 - 1}\right)\omega_n e^{\left(-\zeta + \sqrt{\zeta^2 - 1}\right)\omega_n t} + B\left(-\zeta - \sqrt{\zeta^2 - 1}\right)\omega_n e^{\left(-\zeta - \sqrt{\zeta^2 - 1}\right)\omega_n t}$$

B . C. at $t = 0$ $\ x = x(0),\ \dot{x} = \dot{x}(0)$

B. C (1) $\rightarrow x(0) = A + B$ $\rightarrow A = x(0) - B$

B. C. (2) $\rightarrow \dot{x}(0) = A\left(-\zeta + \sqrt{\zeta^2 - 1}\right)\omega_n + B\left(-\zeta - \sqrt{\zeta^2 - 1}\right)\omega_n$

$$\dot{x}(0) = \{x(0) - B\}\left(-\zeta + \sqrt{\zeta^2 - 1}\right)\omega_n + B\left(-\zeta - \sqrt{\zeta^2 - 1}\right)\omega_n$$

$$\Rightarrow \dot{x}(0) + x(0)\left(-\zeta - \sqrt{\zeta^2 - 1}\right)\omega_n = B\left\{+\zeta - \sqrt{\zeta^2 - 1} - \zeta - \sqrt{\zeta^2 - 1}\right\}$$

$$B = \frac{-\dot{x}(0) - x(0)\left(-\zeta - \sqrt{\zeta^2 - 1}\right)\omega_n}{2\sqrt{\zeta^2 - 1}\,\omega_n}$$

$$A = x(0) - B$$

$$= x(0) - \frac{-\dot{x}(0) - x(0)\left(-\zeta - \sqrt{\zeta^2 - 1}\right)\omega_n}{2\sqrt{\zeta^2 - 1}\,\omega_n}$$

$$= \frac{\dot{x}(0) - x(0)\left(\zeta + \sqrt{\zeta^2 - 1}\right)\omega_n}{2\omega_n\sqrt{\zeta^2 - 1}}$$

$x = Ae^{\left(-\zeta + \sqrt{\zeta^2 - 1}\right)\omega_n t} + Be^{\left(-\zeta - \sqrt{\zeta^2 - 1}\right)\omega_n t} \rightarrow$ Response equation

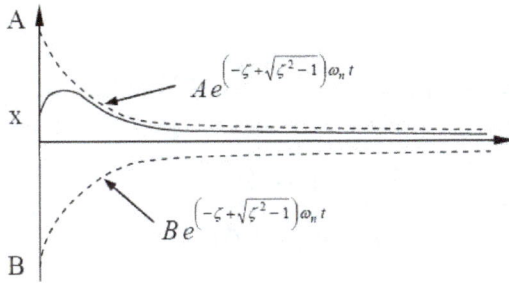

Oscillatory motion ($\zeta > 1.0$)

Special Case (from General Damped Solution)

Free undamped vibration

$$c = 0$$

$$x = A_1 \sin \omega_n t + A_2 \cos \omega_n t$$

Where $\omega_n = \sqrt{k/m} \rightarrow$ natural frequency

$$x = A \sin\left(\omega_n t + \phi\right)$$

To determine A_1, A_2 and A we have to apply the initial condition at $t = 0$.

at $t = 0$ $x = x_0$, $\dot{x} = v_0$

$$x(t) = x_0 \cos \omega_n t + \frac{v_0}{\omega_n} \sin \omega_n t$$

Logarithmic Decrement from Damped Free Vibration

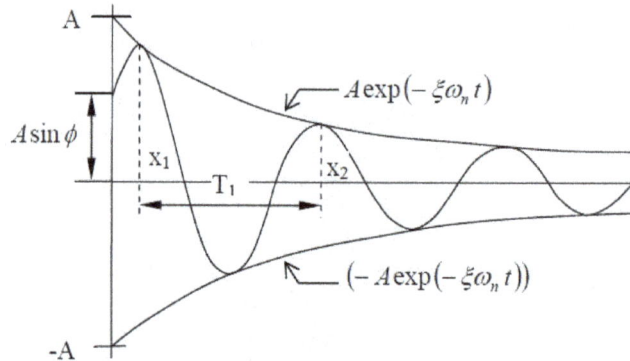

Height of two consecutive picks x_1, x_2. We can define a quantity called Logarithmic Decrement

$$\delta = \ln \frac{x_1}{x_2} = \ln \frac{A e^{-\xi \omega_n t} \sin(\omega_d t + \phi)}{A e^{-\xi \omega_n (t+T)} \sin\left[\omega_d (t+T) + \phi\right]}$$

$$= \ln \frac{e^{-\xi \omega_n t} \sin(\omega_d t + \phi)}{e^{-\xi \omega_n (t+T)} \sin(\omega_d t + \phi)} \qquad T = \frac{2\pi}{\omega_d}$$

$$\sin\left[\omega_d \left(t + \frac{2\pi}{\omega_d}\right) + \phi\right] = \sin\left(\omega_d + 2\pi + \phi\right) = \sin\left(\omega_d t + \phi\right)$$

$$\delta = \xi \omega_n T = \xi \omega_n \frac{2\pi}{\omega_d}$$

$$= \xi \omega_n = \frac{2\pi}{\omega_n \sqrt{1 - \xi^2}} \qquad \xi \rightarrow \text{very small}$$

$$\delta = 2\pi\xi \implies \xi = \frac{\delta}{2\pi}$$

The damping factor ξ can also be determined by measuring two displacements separated by any number of complete cycles. Letting x_1 and $x_j + 1$ be the amplitudes corresponding to the times t_1 and $t_{j+1} = t_1 + j * T$, where j is an integer, we conclude that

$$\frac{x_1}{x_{j+1}} = \frac{x_1}{x_2} \cdot \frac{x_2}{x_3} \cdot \frac{x_3}{x_4} \ldots\ldots\ldots \frac{x_j}{x_{j+1}} = e^{(\xi\omega_n T)j}$$

$$\delta = \frac{1}{j} \ln \frac{x_1}{x_{j+1}} = \frac{1}{j} \ln \left(e^{(\xi\omega_n T)j}\right) = \xi\omega_n T$$

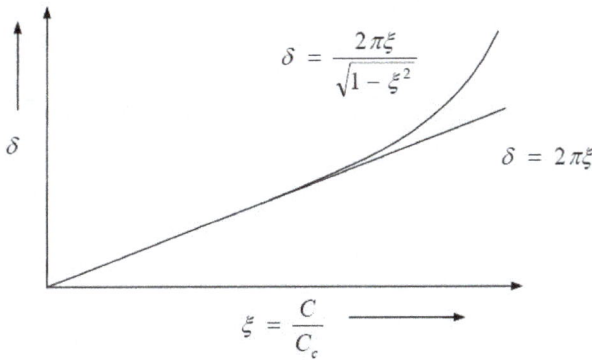

Graphical representation of linear and non-linear logarithmic decrement curves

Summary of Expressions

$$\lambda_{1,2} = \left(-\xi \pm \sqrt{\xi^2 - 1}\right)\omega_n$$

when $\xi > 1 \rightarrow \lambda_1$ & λ_2 are real

$$x = A_1 \exp\left(-\xi + \sqrt{\xi^2 - 1}\right)\omega_n t + A_2 \exp\left(-\xi - \sqrt{\xi^2 - 1}\right)\omega_n t$$

When $\xi = 1$ $\qquad \lambda_{1,2} = -\omega_n$

$$x = \left(A_1 + A_2 t\right)e^{-\omega_n t}$$

When $\xi < 1$, $\qquad \lambda_{1,2}$ are complex

$$x = A_1 \exp\left(-\xi + i\sqrt{\xi^2 - 1}\right)\omega_n t + A_2 \exp\left(-\xi - i\sqrt{\xi^2 - 1}\right)\omega_n t$$

$$\omega_d = \omega_n \sqrt{1 - \xi^2}$$

$$\delta = \ln\frac{x_1}{x_2} = \frac{2\pi\xi}{\sqrt{1 - \xi^2}}$$

For small damping, $\delta = 2\pi\xi$

> When damping is very small you should take a series of peaks (2 peaks will give erroneous result)
>
> $$\delta = \frac{1}{n} \ln \frac{x_1}{x_{n+1}}$$
>
> n = no of peaks

Numerical Example

Problem: Initial condition and damping conditions are given as follows

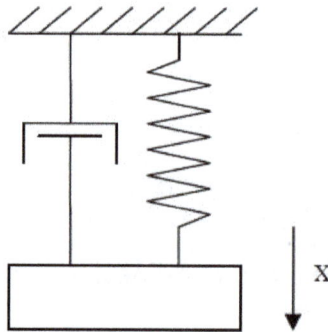

$$t = 0; \quad x = 0; \quad \dot{x} = v_0$$

$$\xi = 2.5, 1, 0.1$$

Find out the response

Solution

There are three cases

1) Over damped

2) Critically damped

3) Under damped

Case I $\xi = 2.5$

$$x = A_1 \exp\left(-\xi + \sqrt{\xi^2 - 1}\right)\omega_n t + A_2 \exp\left(-\xi - \sqrt{\xi^2 - 1}\right)\omega_n t$$

$$x = A_1 e^{-0.209\omega_n t} + A_2 e^{-4.79\omega_n t}$$

To find A_1, A_2 use initial condition

$$\Rightarrow A_1 = -A_2 = \frac{v_0}{\omega_n}(0.218)$$

$$x = 0.218\frac{v_0}{\omega_n}\left(e^{-0.209\omega_n t} + e^{-4.79\omega_n t}\right)$$

Case II

$$\xi = 1$$
$$x = \left(A_1 + A_2 t\right)e^{-\omega_n t}$$

Appling B. C. $\Rightarrow A_1 = 0; A_2 = v_0$

$$x = V_0 t e^{\omega_n t}$$

Case III

$$\xi = 0.1$$

$$x = e^{-0.1\omega_n t}\left(A_1 \sin 0.995 \,\omega_n t + B \cos 0.995 \,\omega_n t\right)$$

From the initial conditions

$$x = 0 \rightarrow B = 0$$

Coulomb Damping

Coulomb Damping is a type of constant mechanical damping in which energy is absorbed via sliding friction. The friction generated by the relative motion of the two surfaces that press against each other is a source of energy dissipation. In general, damping is the dissipation of energy from a vibrating system where the kinetic energy is converted into heat by the friction. Coulomb damping is a common damping mechanism that occurs in machinery.

History

Coulomb damping was so named because Charles-Augustin de Coulomb carried on research in mechanics. He later published a work on friction in 1781 entitled "Theory of Simple Machines" for an Academy of Sciences contest. Coulomb then gained much fame for his work with electricity and magnetism.

Modes of Coulomb Damping

Coulomb damping absorbs energy with friction, which converts that kinetic energy into thermal energy or heat. The Coulomb friction law is associated with two aspects. Static and kinetic frictions occur in a vibrating system undergoing Coulomb damping. Static friction occurs when the two objects are stationary or undergoing no relative motion. For static friction, the friction force F exerted between the surfaces having no relative motion cannot exceed a value that is proportional to the product of the normal force N and the coefficient of static friction μ_s,

$$F_s = \mu_s N.$$

Kinetic friction occurs when the two objects are undergoing relative motion and they are sliding against each other. The friction force F exerted between the moving surfaces is equal to a value that is proportional to the product of the normal force N and the coefficient of kinetic friction μ_k,

$$F_k = \mu_k N.$$

In both of these cases, the frictional force always opposes the direction of motion of the object. The normal force is perpendicular to the direction of motion of the object and equal to the weight of the object sliding.

Example

For a simple example, a block of mass m slides over a rough horizontal surface under the restraint of a spring with a spring constant k. The spring is attached to the block and mounted to an immobile object on the other end allowing the block to be moved by the force of the spring,

$$F = kx.$$

Because the surface is horizontal, the normal force is constant and equal to the weight of the block, or N=mg. This can be determined by summing the forces in the vertical direction. A position x is then measured horizontally to the right from the location of the block when the spring is unstretched. As stated earlier, the friction force acts in a direction opposite the motion of the block. Once put into motion the block will oscillate back and forth around the equilibrium position. Newton's Second Law states that the equation of motion of the block is $m\ddot{x} = -kx - F$ or $m\ddot{x} = -kx + F$ depending on the direction of motion of the block. In this equation $\ddot{x}$ is the acceleration of the block and x is the position of the block. A real-life example of Coulomb damping occurs in large structures with non-welded joints such as airplane wings.

Theory

Coulomb damping dissipates energy constantly because of sliding friction. The magnitude of sliding friction is a constant value; independent of surface area, displacement or position, and velocity. The system undergoing Coulomb damping is periodic or oscillating and restrained by the sliding friction. Essentially, the object in the system is vibrating back and forth around an equilibrium point. A system being acted upon by Coulomb damping is nonlinear because the frictional force always opposes the direction of motion of the system as stated earlier. And because there is friction present, the amplitude of the motion decreases or decays with time. Under the influence of Coulomb damping, the amplitude decays linearly with a slope of $\pm((2\mu mg\omega_n)/(\pi k))$ where ω_n is the natural frequency. The natural frequency is the number of times the system oscillates between a fixed time interval in an undamped system. It should also be known that the frequency and the period of vibration do not change when the damping is constant, as in the case of Coulomb damping. The period τ is the amount of time between the repetition of phases during vibration. As time progresses, the object sliding slows and the distance it travels during these oscillations becomes smaller until it reaches zero, the equilibrium point. The position where the object stops, or its equilibrium position, could potentially be at a completely different position than when initially at rest because the system is nonlinear. Linear systems have only a single equilibrium point.

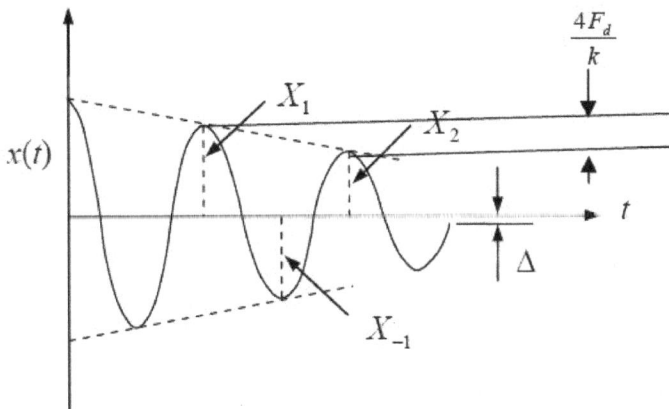

Coulomb Damping results from the sliding of two dry surfaces. The damping force is equal to the product of the normal forces and the coefficient of friction μ and is assumed to be independent of the velocity, once the motion is initiated. Since the sign of the damping force is opposite to that of velocity, the differential equation of motion for each sign is valid only for half cycle intervals.

To determine the decay of amplitude, we resort to the work-energy principle of equating the work done to the change in kinetic energy. Choosing a half-cycle staring at the extreme position with velocity equal to zero and the amplitude equal to X1, the change in kinetic energy is zero and the work done on m is also zero.

$$\tfrac{1}{2}k\left(X_1^2 - X_{-1}^2\right) - F_d\left(X_1 + X_{-1}\right) = 0$$

or, $\tfrac{1}{2}k\left(X_1 - X_{-1}\right) - F_d = 0 \Rightarrow X_1 - X_{-1} = \dfrac{2F_d}{K}$

Where X_{-1} is the amplitude after half-cycle.

Repeating this procedure for the next half-cycle, a further decrease in amplitude of $2F_d / K$ will be found. So that decay in amplitude per cycle is a constant and equal to:

$$X_1 - X_2 = \frac{4F_d}{k}$$

The motion will cease, however, when the amplitude is less than Δ, at which position the spring force is insufficient to over come static friction force, which is generally greater than the kinetic friction force.

Coulomb Damping: Dry Friction

Coulomb damping arises when the oscillating bodies are on the dry surface. For motion to begin, there must be a force acting upon the body that over comes the resistance to motion caused by friction.

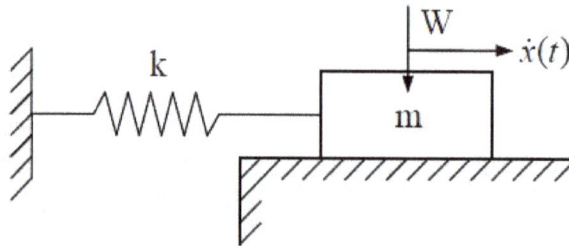

Denoting by F_d the magnitude of the damping force, where $F_d = \mu_k{}^w$, the equation of motion can be written in the form:

$$m\ddot{x} + F_d \, \text{sgn}\,(\dot{x}) + k\,x \;=\; 0$$

Where sgn () denotes sign of and represents a function having the value +1 if its arguments $\dot{x}$ is positive and the value −1 if its arguments is negative. Mathematically

$$\text{sgn}\left(\dot{x}\right) = \frac{\dot{x}}{\left|\dot{x}\right|}$$

∴ We can write

$$m\ddot{x} + kx = -F_d \qquad for \ \dot{x} > 0$$
$$m\ddot{x} + kx = F_d \qquad for \ \dot{x} < 0$$

Take $x(0) = x_0$, where the initial displacement x_0 sufficiently large that the restoring force in the spring exceed the static friction force. Because in the ensuing motion the velocity is negative, we must solve second equation first and the solution can be obtained as

a) Complementary solution and

b) Particular solution

$$\left(a\right)\ddot{x} + \omega_n^2 = 0 \rightarrow x\left(t\right) = \frac{v_1}{\omega_n}\sin \omega_n t + x_1 \cos \omega_n t$$

Note: Solution of undamped SDOF with initial displacement $\left(x_1\right)$ and velocity $\left(v_1\right)$

where $\omega_n^2 = \dfrac{k}{m}$

$$\left(b\right)\ddot{x} + \omega_n^2 x = \omega_n^2 f_d$$

In which $f_d = F_d / k$ represent an equivalent displacement. Now we can solve the above equation with initial conditions as $x(0) = x_0$ & $\dot{x}(0) = 0$ and the solution is

$$x\left(t\right) = \left(x_0 - f_d\right)\cos \omega_n t + f_d$$

Differentiating w.r.t.t we get,

$$\dot{x}\left(t\right) = -\omega_n \left(x_0 - f_d\right)\sin \omega_n t$$

the condition $\dot{x}\left(t\right) = 0$ leads to $t_1 = \dfrac{\pi}{\omega_n}$ at which the displacement is $x\left(t_1\right) = -x_0 + 2f_d$.
If $x\left(t_1\right)$ is sufficiently large in magnitude to overcome the static friction, then the mass acquires a positive velocity, so that the motion must satisfy the equation

$$\ddot{x} + \omega_n^2 x = -\omega_n^2 f_d$$

Where $x(t)$ is subject to initial condition $x(t_1) = -x_0 + 2f_d$, $\dot{x}(t_1) = 0$. The solution is

$$x\left(t\right) = \left(x_0 - 3f_d\right)\cos \omega_n t - f_d$$

Solution is valid in the interval $t_1 \leq t \leq t_2$ where t_2 is the next value of time in which the velocity reduces to zero. This value is $t_2 = \dfrac{2\pi}{\omega_n}$ at which time the velocity is ready to reverse direction once again, this time from right to left. The displacement at $t = t_2$ is

$$x(t_2) = x_0 - 4f_d$$

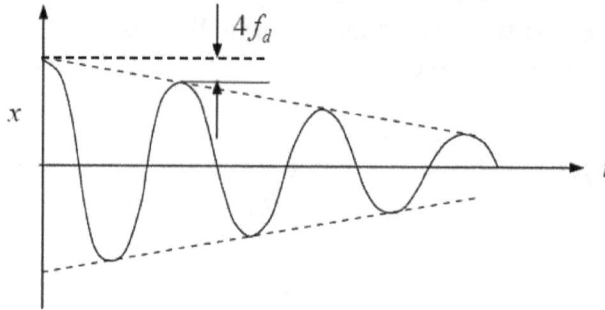

Complementary Solution

$$\ddot{x} + \omega_n^2 x = 0$$

$$x(t) = x_1 \cos \omega_n t + \frac{v_1}{\omega_n} \sin \omega_n t$$

Particular Solution

$$\ddot{x} + \omega_n^2 x = \omega_n^2 f_d$$

$$\Rightarrow \left(D^2 + \omega_n^2\right) x = \omega_n^2 f_d$$

$$\Rightarrow x = \frac{\omega_n^2 f_d}{D^2 + \omega_n^2}$$

$$= \frac{\omega_n^2 f_d}{\omega_n^2 \left(1 + \dfrac{D^2}{\omega_n^2}\right)}$$

$$= \left(1 + \frac{D^2}{\omega_n^2}\right)^{-1} f_d$$

$$\approx \left(1 - \frac{D^2}{\omega_n^2}\right) f_d$$

$$= f_d$$

$\therefore$ Total Solution:

$$x(t) = x_1 \cos \omega_n t + \frac{v_1}{\omega_n} \sin \omega_n t + f_d$$

$$\dot{x}(t) = -\omega_n x_1 \sin \omega_n t + v_1 \cos \omega_n t$$

$$t = 0 \qquad\qquad x_0 = x_1 + f_d \Rightarrow x_1 = (x_0 - f_d)$$

$$x(t) = (x_0 - f_d) \cos \omega_n t + f_d$$

Avionics

Radar and other avionics in the nose of a Cessna Citation I/SP

Avionics are the electronic systems used on aircraft, artificial satellites, and spacecraft. Avionic systems include communications, navigation, the display and management of multiple systems, and the hundreds of systems that are fitted to aircraft to perform individual functions. These can be as simple as a searchlight for a police helicopter or as complicated as the tactical system for an airborne early warning platform. The term *avionics* is a portmanteau of the words *aviation* and *electronics*.

F105 Thunderchief with avionics laid out

History

The term "avionics" was coined by the journalist Philip J. Klass as a portmanteau of "aviation electronics". Many modern avionics have their origins in World War II wartime developments. For example, autopilot systems that are prolific today were started to help bomber planes fly steadily enough to hit precision targets from high altitudes. Famously, radar was developed in the UK, Germany, and the United States during the same period. Modern avionics is a substantial portion of military aircraft spending. Aircraft like the F15E and the now retired F14 have roughly 20 percent of their budget spent on avionics. Most modern helicopters now have budget splits of 60/40 in favour of avionics.

The civilian market has also seen a growth in cost of avionics. Flight control systems (fly-by-wire) and new navigation needs brought on by tighter airspaces, have pushed up development costs. The major change has been the recent boom in consumer flying. As more people begin to use planes as their primary method of transportation, more elaborate methods of controlling aircraft safely in these high restrictive airspaces have been invented.

Modern Avionics

Avionics plays a heavy role in modernization initiatives like the Federal Aviation Administration's (FAA) Next Generation Air Transportation System project in the United States and the Single European Sky ATM Research (SESAR) initiative in Europe. The Joint Planning and Development Office put forth a roadmap for avionics in six areas:

- Published Routes and Procedures – Improved navigation and routing

- Negotiated Trajectories – Adding data communications to create preferred routes dynamically

- Delegated Separation – Enhanced situational awareness in the air and on the ground

- LowVisibility/CeilingApproach/Departure – Allowing operations with weather constraints with less ground infrastructure

- Surface Operations – To increase safety in approach and departure

- ATM Efficiencies – Improving the ATM process

Aircraft Avionics

The cockpit of an aircraft is a typical location for avionic equipment, including control, monitoring, communication, navigation, weather, and anti-collision systems. The majority of aircraft power their avionics using 14- or 28volt DC electrical systems; however, larger, more sophisticated aircraft (such as airliners or military combat aircraft) have AC systems operating at 400 Hz, 115 volts AC. There are several major vendors of flight avionics, including Panasonic Avionics Corporation, Honeywell (which now owns Bendix/King), Rockwell Collins, Thales Group, GE Aviation Systems, Garmin, Parker Hannifin, UTC Aerospace Systems and Avidyne Corporation.

One source of international standards for avionics equipment are prepared by the Airlines Electronic Engineering Committee (AEEC) and published by ARINC.

Communications

Communications connect the flight deck to the ground and the flight deck to the passengers. Onboard communications are provided by public-address systems and aircraft intercoms.

The VHF aviation communication system works on the airband of 118.000 MHz to 136.975 MHz. Each channel is spaced from the adjacent ones by 8.33 kHz in Europe, 25 kHz elsewhere. VHF is also used for line of sight communication such as aircraft-to-aircraft and aircraft-to-ATC. Amplitude modulation (AM) is used, and the conversation is performed in simplex mode. Aircraft communication can also take place using HF (especially for trans-oceanic flights) or satellite communication.

Navigation

Navigation is the determination of position and direction on or above the surface of the Earth. Avionics can use satellite-based systems (such as GPS and WAAS), ground-based systems (such as VOR or LORAN), or any combination thereof. Navigation systems calculate the position automatically and display it to the flight crew on moving map displays. Older avionics required a pilot or navigator to plot the intersection of signals on a paper map to determine an aircraft's location; modern systems calculate the position automatically and display it to the flight crew on moving map displays.

Monitoring

The Airbus A380 glass cockpit featuring pull-out keyboards and two wide computer screens on the sides for pilots.

The first hints of glass cockpits emerged in the 1970s when flight-worthy cathode ray tube (CRT) screens began to replace electromechanical displays, gauges and instruments. A "glass" cockpit refers to the use of computer monitors instead of gauges and other analog displays. Aircraft were getting progressively more displays, dials and information dashboards that eventually competed for space and pilot attention. In the 1970s, the average aircraft had more than 100 cockpit instruments and controls.

Glass cockpits started to come into being with the Gulfstream GIV private jet in 1985. One of the key challenges in glass cockpits is to balance how much control is automated and how much the pilot should do manually. Generally they try to automate flight operations while keeping the pilot constantly informed.

Aircraft Flight-control System

Aircraft have means of automatically controlling flight. Autopilot was first invented by Lawrence Sperry during World War I to fly bomber planes steady enough to hit accurate targets from 25,000 feet. When it was first adopted by the U.S. military, a Honeywell engineer sat in the back seat with bolt cutters to disconnect the autopilot in case of emergency. Nowadays most commercial planes are equipped with aircraft flight control systems in order to reduce pilot error and workload at landing or takeoff.

The first simple commercial auto-pilots were used to control heading and altitude and had limited authority on things like thrust and flight control surfaces. In helicopters, auto-stabilization was used in a similar way. The first systems were electromechanical. The advent of fly by wire and electro-actuated flight surfaces (rather than the traditional hydraulic) has increased safety. As with displays and instruments, critical devices that were electro-mechanical had a finite life. With safety critical systems, the software is very strictly tested.

Collision-avoidance Systems

To supplement air traffic control, most large transport aircraft and many smaller ones use a traffic alert and collision avoidance system (TCAS), which can detect the location of nearby aircraft, and provide instructions for avoiding a midair collision. Smaller aircraft may use simpler traffic alerting systems such as TPAS, which are passive (they do not actively interrogate the transponders of other aircraft) and do not provide advisories for conflict resolution.

To help avoid controlled flight into terrain (CFIT), aircraft use systems such as ground-proximity warning systems (GPWS), which use radar altimeters as a key element. One of the major weaknesses of GPWS is the lack of "look-ahead" information, because it only provides altitude above terrain "look-down". In order to overcome this weakness, modern aircraft use a terrain awareness warning system (TAWS).

Black Boxes

Commercial aircraft cockpit data recorders, commonly known as a "black box", store flight information and audio from the cockpit. They are often recovered from a plane after a crash to determine control settings and other parameters during the incident.

Weather Systems

Weather systems such as weather radar (typically Arinc 708 on commercial aircraft) and lightning detectors are important for aircraft flying at night or in instrument meteorological conditions, where it is not possible for pilots to see the weather ahead. Heavy precipitation (as sensed by radar) or severe turbulence (as sensed by lightning activity) are both indications of strong convective activity and severe turbulence, and weather systems allow pilots to deviate around these areas.

Lightning detectors like the Stormscope or Strikefinder have become inexpensive enough that they are practical for light aircraft. In addition to radar and lightning detection, observations and extended radar pictures (such as NEXRAD) are now available through satellite data connections, allowing pilots to see weather conditions far beyond the range of their own in-flight systems. Modern displays allow weather information to be integrated with moving maps, terrain, and traffic onto a single screen, greatly simplifying navigation.

Modern weather systems also include wind shear and turbulence detection and terrain and traffic warning systems. Inplane weather avionics are especially popular in Africa, India, and other countries where air-travel is a growing market, but ground support is not as well developed.

Aircraft Management Systems

There has been a progression towards centralized control of the multiple complex systems fitted to aircraft, including engine monitoring and management. Health and usage monitoring systems (HUMS) are integrated with aircraft management computers to give maintainers early warnings of parts that will need replacement.

The integrated modular avionics concept proposes an integrated architecture with application software portable across an assembly of common hardware modules. It has been used in fourth generation jet fighters and the latest generation of airliners.

Mission or Tactical Avionics

Military aircraft have been designed either to deliver a weapon or to be the eyes and ears of other weapon systems. The vast array of sensors available to the military is used for whatever tactical means required. As with aircraft management, the bigger sensor platforms (like the E3D, JSTARS, ASTOR, Nimrod MRA4, Merlin HM Mk 1) have mission-management computers.

Police and EMS aircraft also carry sophisticated tactical sensors.

Military Communications

While aircraft communications provide the backbone for safe flight, the tactical systems are designed to withstand the rigors of the battle field. UHF, VHF Tactical (30–88 MHz) and SatCom systems combined with ECCM methods, and cryptography secure the communications. Data links such as Link 11, 16, 22 and BOWMAN, JTRS and even TETRA provide the means of transmitting data (such as images, targeting information etc.).

Radar

Airborne radar was one of the first tactical sensors. The benefit of altitude providing range has meant a significant focus on airborne radar technologies. Radars include airborne early warning (AEW), anti-submarine warfare (ASW), and even weather radar (Arinc 708) and ground tracking/proximity radar.

The military uses radar in fast jets to help pilots fly at low levels. While the civil market has had weather radar for a while, there are strict rules about using it to navigate the aircraft.

Sonar

Dipping sonar fitted to a range of military helicopters allows the helicopter to protect shipping assets from submarines or surface threats. Maritime support aircraft can drop active and passive sonar devices (sonobuoys) and these are also used to determine the location of hostile submarines.

Electro-Optics

Electro-optic systems include devices such as the head-up display (HUD), forward looking infrared (FLIR), infra-red search and track and other passive infrared devices (Passive infrared sensor). These are all used to provide imagery and information to the flight crew. This imagery is used for everything from search and rescue to navigational aids and target acquisition.

ESM/DAS

Electronic support measures and defensive aids are used extensively to gather information about threats or possible threats. They can be used to launch devices (in some cases automatically) to counter direct threats against the aircraft. They are also used to determine the state of a threat and identify it.

Aircraft Networks

The avionics systems in military, commercial and advanced models of civilian aircraft are interconnected using an avionics databus. Common avionics databus protocols, with their primary application, include:

- Aircraft Data Network (ADN): Ethernet derivative for Commercial Aircraft

- Avionics Full-Duplex Switched Ethernet (AFDX): Specific implementation of ARINC 664 (ADN) for Commercial Aircraft

- ARINC 429: Generic Medium-Speed Data Sharing for Private and Commercial Aircraft

- ARINC 664

- ARINC 629: Commercial Aircraft (Boeing 777)

- ARINC 708: Weather Radar for Commercial Aircraft

- ARINC 717: Flight Data Recorder for Commercial Aircraft

- ARINC 825: CAN bus for commercial aircraft (for example Boeing 787 and Airbus A350)

- IEEE 1394b: Military Aircraft

- MIL-STD-1553: Military Aircraft

- MIL-STD-1760: Military Aircraft

- TTP – Time-Triggered Protocol: Boeing 787 Dreamliner, Airbus A380, Fly-By-Wire Actuation Platforms from Parker Aerospace

- TTEthernet – Time-Triggered Ethernet: NASA Orion Spacecraft

Aerospace Structural Dynamics

We assume the aircraft to be a rigid body for the study of aircraft stability and control.

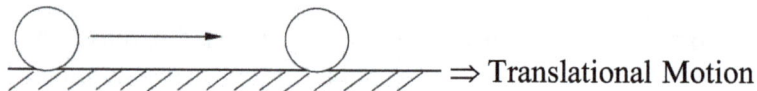

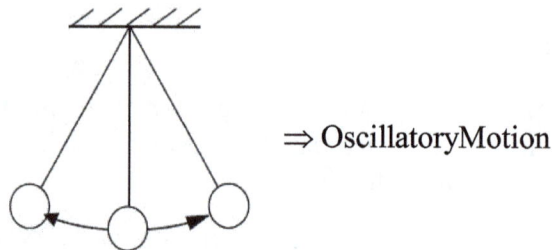

$\Rightarrow$ Translational Motion

$\Rightarrow$ OscillatoryMotion

In case of vibration $\rightarrow$ 3 types of motion

1. Rigid body motion

2. Elastic deformation

3. Periodic motion

$\Rightarrow$ Rigid body motion

Deformation

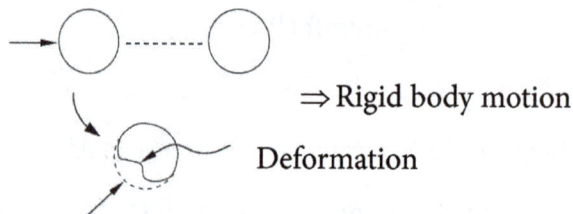

– Based on this deformation, we design the ball.

- In vibration there has to be oscillation as well $\Rightarrow$ Rigid body motion + elastic deformation + oscillatory motion $\rightarrow$ VIBRATION.

Consequences of Vibration

- Passenger and Crew comfort (ideal zero vibration)

- Vibration gives rise to noise $\rightarrow$ Undesirable

- Change of structural integrity

- Reliability of vibration sensitive equipments

1) Discomfort depends on:

a) Frequency

b) Amplitude

c) Duration and

d) Direction

> 4 Hz vertical vibration

Most Sensitive Range $\rightarrow$ 5 – 15 Hz

< 4 Hz Horizontal Vibration

Sl. No	Subjective Rating	Peak Velocity (m/sec) (RMS)	g at 16 Hz (RMS)
1	Comfortable	0.17	0.045
2	A little comfort	0.17 – 0.35	0.045 – 0.090
3	Fairly comfort	0.28 – 0.55	0.072 – 0.144
4	Uncomfortable	0.45 – 0.90	0.115 – 0.280
5	Very uncomfortable	0.70 – 1.45	0.18 – 0.36
6	Extremely uncomfortable	0.70 – 1.45	0.39

2) Structural Integrity

For the structure not to be in total failure (Some level of failure can be present)

$\rightarrow$ Magnification of stresses σ_x or σ_y

$\rightarrow$ Structural fatigue

$\rightarrow$ Large deflection

3) Equipment Reliability

You cannot write the equation unless there is equilibrium in the dynamic system.

D'Alembert's Principle (1743)

In the case of static elastic deformation, the internal elastic forces, try to restore the system to the position of equilibrium.

In the case of pendulum, restoring force is force of gravity (Rigid body motion).

In the case of vibration, D'Alembert reasoned that the sum of the forces acting on a particle result in acceleration $\ddot{x}$ (in the same direction). The inertial force can be computed by multiplying by the mass of the body, $F = m\ddot{x}$. If the force so computed is applied to the body at its mass center in the direction opposite to its acceleration, the dynamic problem is reduced to one of statics. This is defined so-called D'Alembert principle.

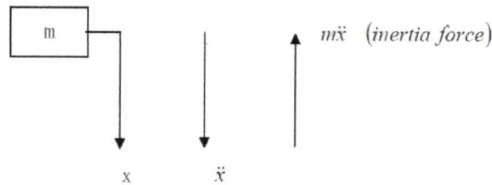

A mechanical system in motion is always in the state of dynamic equilibrium under all forces including the inertia forces.

All forces including motion dependent forces i.e. elastic inertia and damping forces as well as external forces.

References

- Kottasova, Ivana "The world's largest aircraft crashes after 2nd test flight", August 24, 2016, CNN Tech on CNN, the Cable News Network, retrieved November 22, 2016

- Crane, Dale: Dictionary of Aeronautical Terms, third edition, page 194. Aviation Supplies & Academics, 1997. ISBN 1-56027-287-2

- Bergqvist, Pia, "Fastest Airplanes: Top Performers in Their Class," September 17, 2014, Flying Magazine, retrieved December 3, 2016

- Gunston, Bill (1987). Jane's Aerospace Dictionary 1987. London, England: Jane's Publishing Company Limited. ISBN 0-7106-0365-7

- Trujillo, Staff Sgt. Robert M.,"SR-71 Blackbird: Gone but not forgotten," January 26, 2016, 9th Reconnaissance Wing Public Affairs, U.S. Air Force, retrieved December 2, 2016

- "Cessna rolls out first production unit of new Citation X," Apr 15, 2013, Wichita Business Journal, retrieved November 22, 2016

- The Chambers Dictionary. Edinburgh: Chambers Harrap Publishers Ltd. 2000 [1998]. p. 541. ISBN 0-550-14000-X. the gas-bag of a balloon or airship

- Bender, Jeremy and Amanda Macias, "The 9 fastest piloted planes in the world," September 18, 2015, Business Insider, retrieved December 3, 2016

2

Concepts of Aircraft Structural Dynamics

This chapter studies the various motions studied in aircraft structural dynamics. Some of the topics discussed are support motion, rotating unbalance, harmonic motion, etc. The section strategically encompasses and incorporates the major components and key concepts of the subject matter, providing a complete understanding.

Oscillatory Motion

The Study of Vibration is concerned with the oscillatory motions of bodies and the forces associated with them. All bodies possessing mass and elasticity are capable of vibration. Thus most engineering machines and structures experience vibration to some degree, and their design generally requires consideration of oscillatory behavior.

There are two general classes of vibrations – free and forced. Free vibration takes place when a system oscillates under the action of forces inherent in the system itself, and when external impressed forces are absent. The system under free vibration will vibrate at one or more of its natural frequencies, which are properties of the dynamical system established by its mass and stiffness distribution.

Vibration takes place under the excitation of external forces is called forced vibration. When the excitation is oscillatory, the system is forced to vibrate at the excitation frequency. If the frequency of excitation coincides with one of the natural frequencies of the system, a condition of resonance is encountered, and dangerously large oscillations may result. The failure of major structures such as bridges, buildings, or airplane wings is an awesome possibility under resonance. Thus, the calculation of natural frequencies is of major importance in study of vibrations.

Vibrating systems are all subject to damping to some degree because energy is dissipated by friction and other resistances. If the damping is small, it has very little influence on the natural frequencies of the system, and hence the calculations for the natural frequencies are generally made on basis of no damping. On the other hand, damping is of great important in limiting the amplitude of oscillation at resonance. The natural frequencies found in case of an undamped structures is known as the undamped natural frequencies whereas it is known as the damped natural frequencies for a damped structure. The damped natural frequencies are smaller in magnitude compared to its corresponding undamped natural frequencies. One of the most important phenomenon is resonance, where the amplitude of vibration increases to huge magnitude depending upon the presence of damping in the structure. It happens once the forcing frequency matches with any one of the natural frequency of a structure.

The number of independent co ordinates required to describe the motion of a system is called degree of freedom of the system. Thus a free particle undergoing general motion in space will have three degrees of freedom, while a rigid body will have six degrees of freedom, i.e., three components of position and three angles defining its orientation. Furthermore, a continuous elastic body will require an infinite number of coordinates (three for each such point on the body) to describe its motion; hence its degree of freedom must be infinite. However, in many cases, parts of such bodies may be assumed to be rigid, and the system may be considered to be dynamically equivalent to one having finite degrees of freedom. In fact, a surprisingly large number of vibration problems can be treated with sufficient accuracy by reducing the system to one having a few degrees of freedom.

Harmonic Motion

Oscillatory motion may repeat itself regularly, as in the balance wheel of a watch, or display considerable irregularity, as in earthquakes. When the motion is repeated in equal intervals of time τ, it is called periodic motion. The repetition time τ is called period of the oscillation, and its reciprocal, $f = 1/\tau$, is called the frequency. If the motion is designated by the time function $x(t)$, then any periodic motion must satisfy the relationship $x(t) = x(t+\tau)$.

The simplest form of periodic motion is simple harmonic motion. It can be demonstrated by a mass suspended from a light spring. If the mass is displaced from its rest and released, it will oscillate up and down. By placing a light source on the oscillating mass, its motion can be recorded on a light sensitive film strip, which is made to move past it at a constant speed.

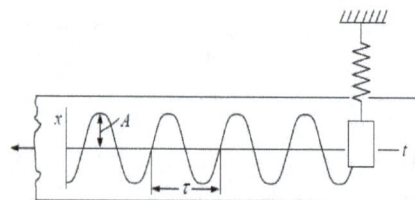

Recording of harmonic motion

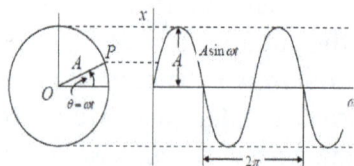

Harmonic motion as projection of a point moving on a circle

The motion recorded on the film strip can be expressed by the equation

$$x = A \sin 2\pi t / \tau$$

where A is the amplitude of the oscillation, measured from the equilibrium position of the mass, and τ is the period. The motion is repeated when $t = \tau$.

Harmonic motion is often represented as the projection on a straight line of a point that is

moving on a circle as constant speed, with the angular speed of the line

OP designated by ω, the displacement x can be written as

$$x = A \sin \omega t$$

The quantity ω is generally measured in radians per second, and is referred to as the *circular frequency*. Since the motion repeats itself in 2π radians, we have the relationship

$$\omega = 2\pi / \tau = 2\pi f$$

where τ and f are the period and frequency of the harmonic motion, usually measured in seconds and cycle per second, respectively.

The velocity and acceleration of the harmonic motion can be simply determined by

differentiation using the dot notation for the derivative, we obtain

$$\dot{x} = \omega A \cos \omega t = \omega A \sin\left(\omega t + \frac{\pi}{2}\right)$$

$$\ddot{x} = -\omega^2 A \sin \omega t = \omega^2 A \sin\left(\omega t + \pi\right)$$

Thus the velocity and acceleration are also harmonic with the same frequency of oscillation but lead the displacement by $\pi/2$ and π radians, respectively.

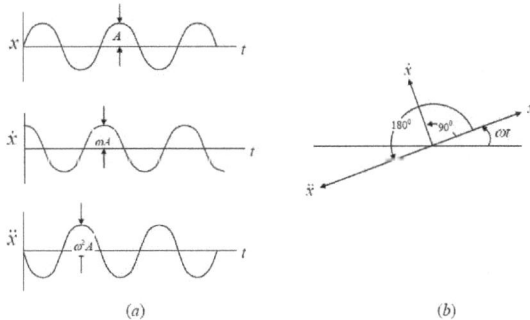

(a)   (b)

In harmonic motion, the velocity and acceleration lead the displacement by $\pi / 2$ *and* π

Above figure shows both time variation and the vector phase relationship between the displacement, velocity, and acceleration in harmonic motion.

$$\ddot{x} = -\omega^2 x$$

Exponential form. The trigonometric functions of sine and cosine are related to the exponential function by Euler's equation.

$$e^{i\theta} = \cos\theta + i\sin\theta$$

A vector of amplitude A rotating at constant angular speed ω can be represented as a complex quantity z in the Argand diagram as shown in the figure.

$$z = Ae^{i\omega t}$$
$$= A\cos\omega t + iA\sin\omega t$$
$$= x + iy$$

The quantity z is referred to as the *complex sinusoid* with x and y as the real and imaginary components. It is evident from this diagram that real component x is expressible in terms of z and z* by the equation.

$$x = \frac{1}{2}(z + z^*) = A\cos\omega t = \mathrm{Re}\, Ae^{i\omega t}$$

where Re stands for the real part of the quantity z. We will find that the exponential form of the harmonic motion often offers mathematical advantages over the trigonometric form.

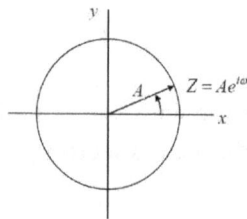

Harmonic motion represented by   Vector z and its conjugate z*.
rotating vector.

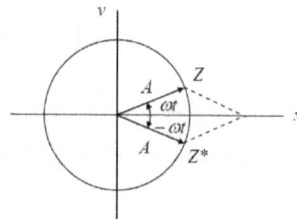

Some of the rules of exponential operations between $z_1 = A_1 e^{i\theta_1}$ and $z_2 = A_2 e^{i\theta_2}$ are:

Multiplication $z_1 z_2 = A_1 A_2 e^{i(\theta_1 + \theta_2)}$

Division $\dfrac{z_1}{z_2} = \left(\dfrac{A_1}{A_2}\right) e^{i(\theta_1 + \theta_2)}$

Powers $z^n = A^n e^{in\theta}$

$$z^{1/n} = A^{1/n} e^{i\theta/n}$$

Periodic Motion

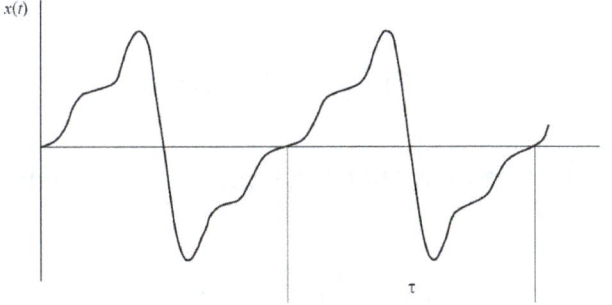

Periodic motion of period τ

The French mathematician J. Fourier (1768-1830) showed that any periodic motion can be represented by a series of sines and cosines that are harmonically related. If $x(t)$ is a periodic function of the period τ, it is represented by the Fourier series

$$x(t) = \frac{a_0}{2} + a_1 \cos \omega_1 t = a_2 \cos \omega_2 t + \ldots\ldots$$
$$+ b_1 \sin \omega_1 t + b_2 \sin \omega_2 t + \ldots\ldots$$

Where

$$\omega_1 = \frac{2\pi}{\tau}$$

$$\omega_n = n\omega_1$$

To determine the coefficients a_n and b_n, we multiply both side by $\cos \omega_n t$ or $\sin \omega_n t$ and integrate each term over the period τ. Recognizing the following relations,

$$\int_{-\tau/2}^{\tau/2} \cos \omega_n t \cos \omega_m t \, dt = \begin{cases} 0 & \text{if } m \neq n \\ \tau/2 & \text{if } m = n \end{cases}$$

$$\int_{-\tau/2}^{\tau/2} \sin \omega_n t \sin \omega_m t \, dt = \begin{cases} 0 & \text{if } m \neq n \\ \tau/2 & \text{if } m = n \end{cases}$$

$$\int_{-\tau/2}^{\tau/2} \cos \omega_n t \sin \omega_m t \, dt = \begin{cases} 0 & \text{if } m \neq n \\ \tau/2 & \text{if } m = n \end{cases}$$

all the terms except one on the right side of the equation will be zero, and we obtain the result

$$a_n = \frac{2}{\tau} \int_{-\tau/2}^{\tau/2} x(t) \cos \omega_n t \, dt$$

$$b_n = \frac{2}{\tau} \int_{-\tau/2}^{\tau/2} x(t) \sin \omega_n t \, dt$$

The Fourier series can also be represented in terms of the exponential function. Substituting

$$\cos \omega_n t = \frac{1}{2} \left(e^{i\omega_n t} + e^{i\omega_n t} \right)$$

$$\sin \omega_n t = \frac{1}{2} \left(e^{i\omega_n t} - e^{i\omega_n t} \right)$$

we obtain

$$x(t) = \frac{a_0}{2} + \sum_{n=1}^{\alpha} \left[\tfrac{1}{2}(a_n - ib) e^{i\omega_n t} + \tfrac{1}{2}(a_n + ib) e^{i\omega_n t} \right]$$

$$= \frac{a_0}{2} + \sum_{n=1}^{\alpha} \left[c_n e^{i\omega_n t} + c_n^* e^{-i\omega_n t} \right]$$

$$= \sum_{n=-\alpha}^{\alpha} c_n e^{i\omega_n t}$$

Where

$$c_0 = \tfrac{1}{2} a_0$$

$$c_n = \tfrac{1}{2}(a_n - ib_n)$$

Substituting for a_n and b_n from above expression we find c_n to be

$$c_n = \tfrac{1}{\tau} \int_{-\tau/2}^{\tau/2} x(t) \left(\cos \omega_n t - i \sin \omega_n t \right) dt$$

$$= \tfrac{1}{\tau} \int_{-\tau/2}^{\tau/2} x(t) e^{-i\omega_n t} \, dt$$

Some computational effort can be minimized when the function $x(t)$ is recognizable in term of the even and odd functions

$$x(t) = E(t) + O(t)$$

An even function $E(t)$ is symmetric about the origin so that $E(t) = E(-t)$, i.e., $\cos \omega t = \cos(-\omega t)$. An odd function satisfies the relationship $O(t) = -O(-t)$, i.e., $\sin \omega t = -\sin(-\omega t)$. The following integrals are then helpful:

$$\int_{-\tau/2}^{\tau/2} E(t) \sin \omega_n t \, dt = 0$$

$$\int_{-\tau/2}^{\tau/2} O(t) \cos \omega_n t \, dt = 0$$

When the coefficients of the Fourier series are plotted against frequency ω_n, the result is a series of discrete lines called the *Fourier Spectrum*. Generally plotted are the absolute values

$$|c_n| = \sqrt{a_n^2 + b_n^2}$$ and the phase $\phi_n = \tan^{-1} b_n / a_n$.

With the aid of the digital computer, harmonics analysis today is efficiently carried out. A computer algorithm known as the *fast Fourier transport* (FFT) is commonly used to minimize the computation time.

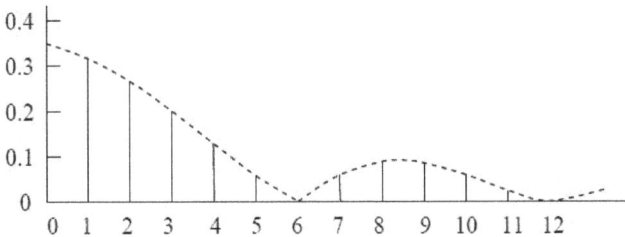

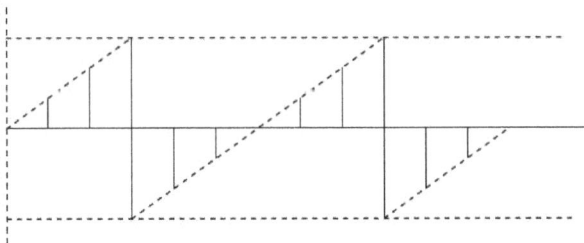

Fourier spectrum for pulses shown in problem may be defined, $k = \dfrac{1}{3}$

Vibration Terminology

The peak value generally indicates the maximum stress that the vibrating part is undergoing. It also places a limitation on the "rattle space" requirement. The term "rattle space" is defined as the available clearance space that permits the system to undergo the induced deflection freely during vibrationis called the rattle space or clearance. If the rattle space is too small to accommodate the deflection of the system,the system will undergo impacts (as it hits the surrounding or nearby surface or object) in each cycle of vibration.

The average value indicates a steady or static value somewhat like the DC level of an electrical current. It can be found by the time integral.

$$\bar{x} = \lim_{T \to \infty} \frac{1}{T} \int_0^T x(t)\, dt$$

For example, the average value for a complete cycle of a sine wave, A sin t, is zero; whereas its average value for a half cycle is

$$\bar{x} = \frac{A}{\pi} \int_0^\pi \sin t\, dt = \frac{2A}{\pi} = 0.637\,A$$

It is evident that this is also the average value of the rectified sine wave shown.

The square of the displacement generally is associated with the energy of the vibration for which the mean square value is a measure. The *mean square value* of a time function *x(t)* is found from the average of the square values, integrated over some time interval T:

$$\overline{x^2} = \lim_{T = \infty} \frac{1}{T} \int_0^T x^2(t)\, dt$$

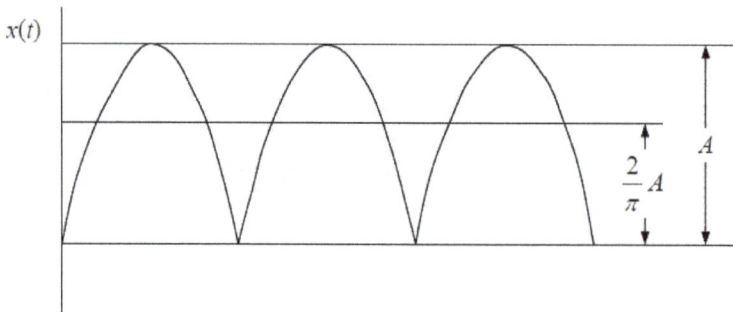

Average value of a rectified sine wave.

For example, if $x(t) = A \sin \omega t$, its mean square value is

$$\overline{x^2} = \lim_{T \to \infty} \frac{A^2}{T} \int_0^T \tfrac{1}{2}(1 - \cos 2\omega t)\, dt = \tfrac{1}{2} A^2$$

The *root mean square* (rms) value is the square root of the mean square value. From the previous example, the rms of the sine wave of amplitude A is $A/\sqrt{2} = 0.707\, A$. Vibrations are commonly measured by rms meters.

Decibel: The decibel is the unit of measurement that is frequently used in vibration measurements. It is define in the terms of power ratio.

$$dB = 10 \log \log_{10} \frac{P_1}{P_2}$$

$$= 10 \log_{10} \left(\frac{x_1}{x_2} \right)^2$$

The second equation results from the fact that power is the proportional to the square of the amplitude of voltage. The decibel is often expressed in term of the first power of amplitude or voltage as

$$Db = 20 \log_{10} \left(\frac{x_1}{x_2} \right)$$

Thus an amplifier with a voltage gain of 5 has a decibel gain of

$$20 \log_{10} (5) = +14$$

Because the decibel is a logarithmic unit, it compresses or expands the scale.

Band	Frequency range (Hz)	Frequency Bandwidth
1	10-20	10
2	20-40	20
3	40-80	40
4	200-400	200

A typical Octave Band

Octave: when the upper limit of a frequency range is twice its lower limit, the frequency span is said to be an *octave*. For example, each of the frequency bands in the figure above represents an octave band.

Harmonic Oscillation

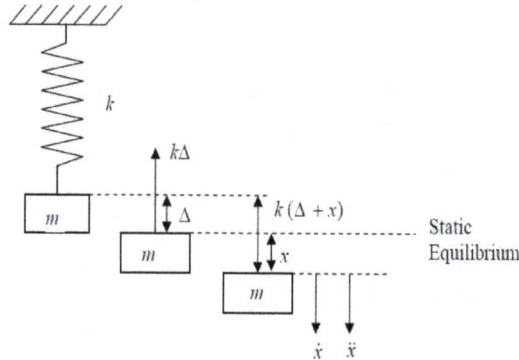

$$kΔ = w = mg$$

$$m\ddot{x} = ΣF = w - k(Δ + x) = -kx$$

$$\Rightarrow m\ddot{x} + kx = 0$$

$$Define, \quad \omega_n^2 = \frac{k}{m}$$

$$\ddot{x} + \omega_n^2 x = 0$$

If we define $x = Ae^{\lambda t}$

$$\lambda^2 + \omega_n^2 = 0 \Rightarrow \lambda = \pm i\omega_n$$

$$x(t) = A_1 e^{i\omega_n t} + A_2 e^{-i\omega_n t}$$

Where A_1 and A_2 are constants of integration. Their values depend on the initial displacement $x(0)$ and initial velocity $\dot{x}(0)$.

Trigonometric Solution

$$x = A \sin \omega_n t + B \cos \omega_n t$$

Where A and B are the two necessary constants depend on initial conditions.

Complex Variable

$x^2 + 1 = 0$ or similar equation $\rightarrow$ no real solution

so, complex numbers are introduced.

$x = \pm\sqrt{-1} = \pm\, i$ where i is called the imaginary unit. So, we can define complex number as $a + ib, c + id$, etc. where a and b etc. are two real number and i is imaginary unit. The absolute value (modulus) of $a + ib$ is defined as $|a + ib| = \sqrt{a^2 + b^2}$. The complex conjugate of $a + ib$ is defined as $a - ib$.

Polar form of Complex Number

Since a complex number $a + ib$ can be considered as an ordered pair (x, y), we can represent such numbers by points in xy plane called complex plane or Argand Diagram.

Argand Diagram

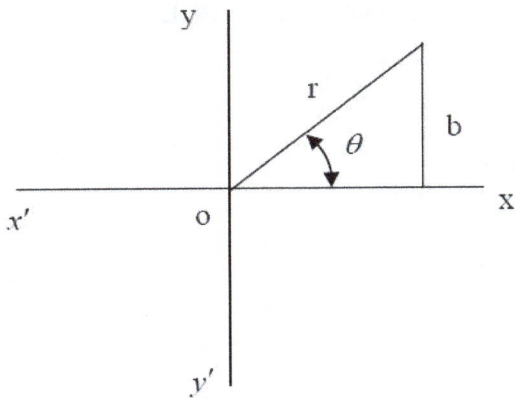

$$a = r\cos\theta$$
$$b = r\sin\theta$$

$$\left.\begin{array}{l} a = r\cos\theta \\ b = r\sin\theta \end{array}\right\} \quad \begin{array}{l} where\, r = \sqrt{a^2 + b^2} \\ = |a + ib| \\ and\ \theta = \tan^{-1}\left(b/a\right) \end{array}$$

$$z = a + ib = r\left(\cos\theta + i\sin\theta\right) \rightarrow cis\,\theta$$

Maclaurin's infinite series can be obtained as

$$\cos\theta = 1 - \frac{\theta^2}{2!} + \frac{\theta^4}{4!} - \ldots\ldots\ldots\ldots$$

$$\sin\theta = \theta - \frac{\theta^3}{3!} + \frac{\theta^5}{5!} - \ldots\ldots\ldots\ldots$$

$$\cos\theta + i\sin\theta = 1 + i\theta - \frac{\theta^2}{2!} - i\frac{\theta^3}{3!} + \frac{\theta^4}{4!} + i\frac{\theta^5}{5!} - \ldots\ldots\ldots$$

$$= e^{i\theta}$$

Solution Single Degree of Freedom System (SDOF)

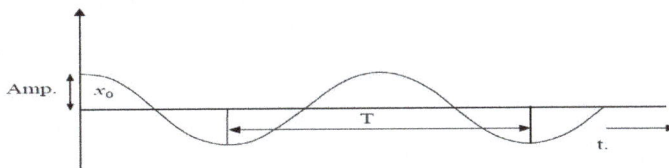

$$\ddot{x} + \omega_n^2 x = 0$$

$$x = A \sin \omega_n t + B \cos \omega_n t$$

$$(i)\, x(0) = x_0 \; , \; (ii)\; \dot{x}(0) = v_0$$

$$(i) \Rightarrow x_0 = B$$

$$(ii) \Rightarrow v_0 = A\omega_n \Rightarrow A = \frac{v_0}{\omega_n}$$

$$\therefore \; x = \frac{v_0}{\omega_n} \sin \omega_n t + x_0 \cos \omega_n t \; \rightarrow \; \text{Trigonometric Solution}$$

Complex Number Solution

$$x = A e^{\lambda t}$$

$$\lambda^2 + \omega_n^2 = 0 \Rightarrow \lambda = \pm i\, \omega_n$$

$$x = A_1 e^{i\omega_n t} + A_2 e^{-i\omega_n t}$$

Where A_1 and A_2 are constant of integration.

Consider the series

$$e^{i\omega t} = \cos \omega t + i \sin \omega t$$

$$e^{-i\omega t} = \cos \omega t - i \sin \omega t$$

Substituting this x reduced to

$$x = (A_1 + A_2) \cos \omega_n t + i(A_1 - A_2) \sin \omega_n t$$

Let, $A_1 + A_2 = A \cos \phi, i(A_1 - A_2) = A \sin \phi$

$$x = A(\cos \phi \cos \omega_n t + \sin \phi \sin \omega_n t)$$

$$A \cos(\omega_n t - \phi)$$

Where new constant of integration are A & ϕ.

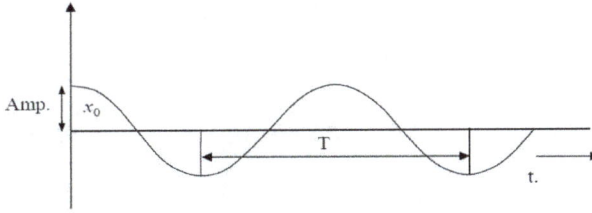

The natural period of oscillation is established from $\omega_n T = 2\pi$

$$T = 2\pi\sqrt{\frac{m}{k}}$$

and the natural frequency is

$$f_n = \frac{1}{T} = \frac{1}{2\pi}\sqrt{\frac{k}{m}} = \frac{\omega_n}{2\pi}$$

Initial condition: $(i)\, x(0) = x_0$ $(ii)\, \dot{x}(0) = v_0$

$$
\left.
\begin{aligned}
(i) &\Rightarrow x_0 = A\cos\phi \\
(ii) &\Rightarrow v_0 = A\omega_n\sin\phi
\end{aligned}
\right\}
\qquad
\begin{aligned}
A &= x_0^2 + \left(\frac{v_0}{\omega_n}\right)^2 \\
\phi &= \tan^{-1}\left(\frac{v_0}{x_0\omega_n}\right)
\end{aligned}
$$

Substituting the constant of integration

$$
\begin{aligned}
x(t) &= A\cos\phi\cos\omega_n t + A\sin\phi\sin\omega_n t \\
&= x_0\cos\omega_n t + \frac{v_0}{\omega_n}\sin\omega_n t
\end{aligned}
\Bigg/
\begin{aligned}
x(t) &= A\cos(\omega_n t - \phi) \\
&= \sqrt{x_0^2 + \left(\frac{v_0}{\omega_n}\right)^2}\cos\left(\omega_n t - \tan^{-1}\frac{v_0}{x_0\omega_n}\right)
\end{aligned}
$$

Characteristic of Harmonic Motion

Units

Force	$N\left(kg\,m/s^2\right)$
Displacement	$x\ (m)$
Acceleration	$\ddot{x}\left(m\ s\right)$
Mass	$m\ (kg)$

Damping coefficient   C (N-s / m)

Velocity $\dot{x}\,(m\,/\,s)$

Linear Stiffness $k\ (N\ /\ m\ \to\ kg - m\ /\ s^2\ /\ m\ \to\ kg\ /\,s^2)$

Angular Velocity   1/s

Moment of Force   N-m

Acceleration due to gravity, $g\ =\ 9.81\ m\ /\ s^2$

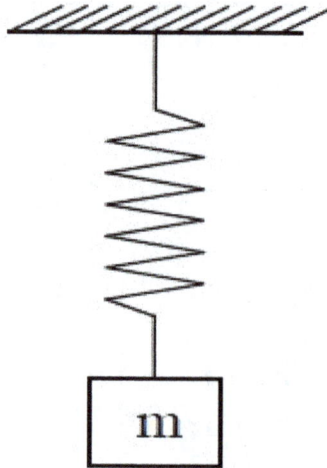

$k = \dfrac{F}{\mu} \leftarrow$ displacement when force F pulls the spring

$$f\,(flexibility) = \frac{1}{k}$$

$\omega_n = \sqrt{\dfrac{k}{m}} \leftarrow$ natural frequency of a single degree freedom system

Characteristic of Harmonic Motion

$$x\ =\ A\sin \omega t$$

$$\dot{x}\ =\ A\omega \cos \omega t$$

$$\omega A \sin\left(\omega t + \tfrac{\pi}{2}\right)$$

$$\ddot{x} = -\omega^2\, A \sin \omega t$$

$$= \omega^2 A \sin\left(\omega t + \pi\right)$$

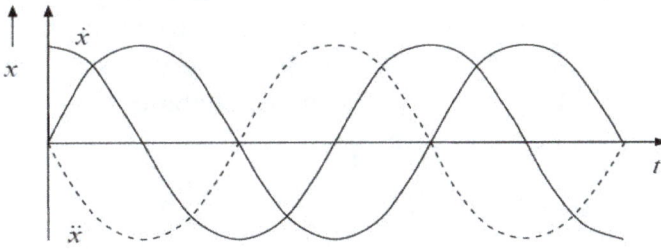

Response diagram of displacement (x), velocity $(\dot{x})$ and acceleration $(\ddot{x})$

Forced Vibration of a Single Degree of Freedom System

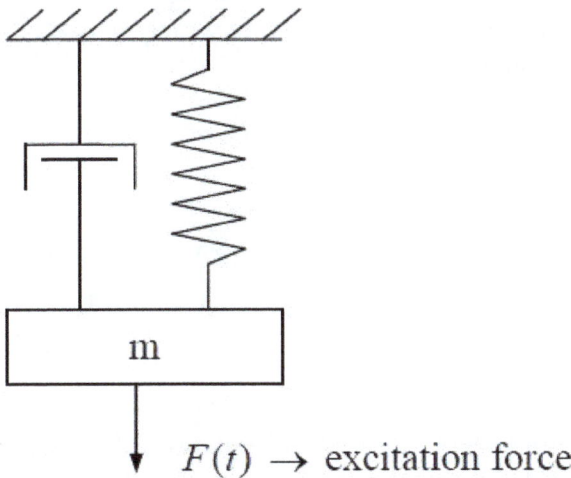

$F(t) \rightarrow$ excitation force

- Excitation force becomes periodic (reciprocating engine).

- Harmonic or periodic forces or rotating or reciprocating parts is exciting in nature.

- Exciting force can also be shock / transient force due to landing of a/c, firing of missile, ejecting of external stores, launch of space vehicle etc.

- Sustained non-periodic random aerodynamic forces experienced by an a/c due to turbulence of the atmosphere.

Response of First - Order Systems to Harmonic Excitation
Frequency Response

The differential equation for a first – order system in the form of a damper – spring system is shown below

$$c\dot{x}(t) + k(x)t = F(t) \qquad\qquad 1$$

where c and k are damping coefficient and stiffness. The homogeneous solution of Eq. (1), can be obtained by letting $F(t) = 0$. In this section we focus our attention on the particular solution, which represents the response to external forces. First, we consider the simplest case, namely, the response to harmonic excitation. To this end, it is convenient to let the force $F(t)$ have the form

$$F(t) = kf(t) = kA\cos\omega t \qquad 2$$

Where ω is the excitation frequency, sometimes referred to as the driving frequency, Note that $f(t)$ and A have units of displacement. The reason for writing the excitation in the form (2) is so as to permit expressing the response in terms of a non-dimensional ratio, as we shall see shortly. No dimensional ratios often enhance the usefulness of a solution by extending its applicability to a large variety of cases. Inserting Eq. (2) into Eq. (1) and dividing through by c, we obtain

$$\dot{x}(t) + ax(t) = Aa\cos\omega t \qquad 3$$

Where

$$a = \frac{k}{c} = \frac{1}{\tau} \qquad 4$$

in which τ is the time constant.

The solution of the homogeneous differential equation, obtained by letting $A = 0$ in Eq. (3), decays exponentially with time, for which reason it is called the transient solution. On the other hand, the particular solution does not vanish as time and is known as the steady – state solution to the harmonic excitation in question. By virtue of the fact that the system is linear, the principle of superposition holds, so that the homogeneous solution and the particular solution can be obtained separately and then combined linearly to obtain the complete solution.

Because the excitation force is harmonic, it can be verified easily that the steady – state response is also harmonic and has same frequenc ω . Moreover, because Eq. (3) involves the function $x(t)$ and its first derivative $\dot{x}(t)$, the response must contain not only $\cos\omega t$ but also $\sin\omega t$. Hence, let us assume that the steady – state solution of Eq. (3) has the form

$$x(t) = C_1\sin\omega t + C_2\cos\omega t \qquad 5$$

Where C_1 and C_2 are constants yet to be determined. Inserting solution (5) into Eq. (3), we obtain

$$\omega(C_1\cos\omega t - C_2\sin\omega t) + a(C_1\sin\omega t + C_2\cos\omega t) = Aa\cos\omega t \qquad 6$$

Equation (6) can be satisfied only if the coefficients of $\sin \omega t$ on the one hand and the coefficients of $\cos \omega t$ on the other hand are the same on both sides of the equation. This, in turn, requires the satisfaction of the equations

$$aC_1 - \omega C_2 = 0$$

$$\omega C_1 + aC_2 = Aa \qquad\qquad 7$$

which represent two algebraic equations in the unknowns C_1 and C_2. Their solution is

$$C_1 = \frac{Aa\omega}{a^2 + \omega^2} \qquad C_2 = \frac{Aa^2}{a^2 + \omega^2} \qquad\qquad 8$$

Introducing Eqs. (8) into Eq. (5), we obtain the steady – state solution

$$x(t) = \frac{Aa}{a^2 + \omega^2}(\omega \sin \omega t + a \cos \omega t) \qquad\qquad 9$$

Solution (9) can be expressed in a more convenient form. To this end, let us introduce the notation

$$\frac{\omega}{\left(a^2 + \omega^2\right)^{1/2}} = \sin \phi \qquad \frac{a}{\left(a^2 + \omega^2\right)^{1/2}} = \cos \omega \qquad\qquad 10$$

Then, Eq. (9) can be written as

$$x(t) = X(\omega)\cos(\omega t - \phi) \qquad\qquad 11$$

Where

$$X(\omega) = \frac{A}{\left[1 + (\omega/a)^2\right]^{1/2}} \qquad\qquad 12$$

Is the amplitude and

$$\phi(\omega) = \tan^{-1}\frac{\omega}{a} \qquad\qquad 13$$

is the phase angle. Both X and ϕ are functions of the excitation frequency ω.

The response to harmonic excitation can be obtained more conveniently by using complex vector representation of the excitation and the response. From Sec. 1.6, we recall that

$$e^{i\omega t} = \cos \omega t + i \sin \omega t \qquad\qquad 14$$

Where $i = \sqrt{-1}$ so that Eq. (2) can be rewritten as

$$F(t) = kf(t) = kA\cos\omega t = \operatorname{Re} kAe^{i\omega t} \qquad 15a$$

Where Re denotes the real part of the function. Similarly in the case of sinusoidal excitation we can write

$$F(t) = kf(t) = kA\sin\omega t = \operatorname{Im} kAe^{i\omega t} \qquad 15b$$

Where Im denotes the imaginary part of the function. Hence, we can rewrite Eq. (3) in the form

$$\dot{x}(t) + ax(t) = aAe^{i\omega t} \qquad 16$$

Then, if the excitation is given by Eq. (15a), we retain the real part of the response and if the excitation is given by Eq. (15b), we retain the imaginary part of the response.

Concentrating once again on the steady – state response, we write the solution of Eq. (16) in the form

$$x(t) = X(i\omega)e^{i\omega t} \qquad 17$$

Inserting Eq. (17) into Eq. (16)

we obtain

$$Z(i\omega)X(i\omega)e^{i\omega t} = aAe^{i\omega t} \qquad 18$$

Where

$$Z(i\omega) = a + i\omega \qquad 19$$

Is the impedance function for this first – order system. Dividing Eq. (18) through by $e^{i\omega t}$ and solving for $X(i\omega)$, we obtain

$$X(i\omega) = \frac{aA}{Z(i\omega)} = \frac{aA}{a+i\omega} = \frac{A}{1+i\omega\tau} \qquad 20$$

Where $\tau = 1/a = c/k$ is the time constant. It will prove convenient to introduce the non dimensional ratio

$$G(i\omega) = \frac{X(i\omega)}{A} = \frac{1}{1+i\omega\tau} = \frac{1-i\omega\tau}{1+(\omega\tau)^2} \qquad 21$$

Where $G(i\omega)$ is known as the frequency response. Inserting Eq. (21) into Eq. (17), we can write the harmonic response in the general form

$$x(t) = AG(i\omega)e^{i\omega t}$$

22

But, the frequency response $G(i\omega)$, as any complex function, can be expressed as

$$G(i\omega) = |G(i\omega)|e^{-i\phi}$$

23

where $|G(i\omega)|$ is the magnitude and ϕ is the phase angle of $G(i\omega)$. Introducing Eq. (23) into Eq. (22), we obtain

$$x(t) = A|G(i\omega)|e^{i(\omega t - \phi)}$$

24

So that if the excitation is in the form of Eq. (15a), the response is the real part of Eq.(24), or

$$x(t) = A|G(i\omega)|\cos(\omega t - \phi)$$

25a

and if the excitation is in the form of Eq. (15b), the response is the imaginery part of Eq. (24), or

$$x(t) = A|G(i\omega)|\sin(\omega t - \phi)$$

25b

From Eqs. (25) it follows that, if the excitation is harmonic with the frequency ω, the response is also harmonic and has the same frequency. Hence, in studying the nature of the response, plotting the response as a function of time will not be very rewarding. Considerably more insight into the system behaviour can be gained by examining how the system responds as the driving frequency ω varies. In particular plots of the magnitude $|G(i\omega)|$ and of the phase angle ϕ versus the frequency ω are very revealing. From complex algebra, if we consider Eq. (21), then we can write

$$|G(i\omega)| = \left[\text{Re}^2\,G(i\omega) + \text{Im}^2\,G(i\omega)\right]^{1/2} = \frac{1}{\left[1 + (\omega\tau)^2\right]^{1/2}}$$

26

And we note from Eq. (12) that $|G(i\omega)| = X(\omega)/A$. The plot $|G(i\omega)|$ versus $\omega\tau$ is shown in the figure. We observe from the figure that for small driving frequencies the magnitude $|G(i\omega)|$ is close to 1 and for high frequencies the magnitude approaches 0. Hence, the system permits low – frequency harmonics to go through undistorted, but it attenuates greatly high – frequency harmonics. For this reason a first – order system is known as a low-pass filter. To obtain the phase angle, we recall first that $e^{-i\phi} = \cos\phi - i\sin\phi$. Then using Eqs. (21) and (23), we can write

$$\phi = \tan^{-1}\left[\frac{-\operatorname{Im}G(i\omega)}{\operatorname{Re}G(i\omega)}\right] = \tan^{-1}\omega\tau$$

27

which checks with Eq. (13). The plot ϕ versus $\omega\tau$ is shown in the figure. The plots $|G(i\omega)|$ versus $\omega\tau$ and ϕ versus $\omega\tau$ are known as frequency – response plots.

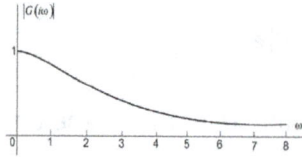

Figure 1

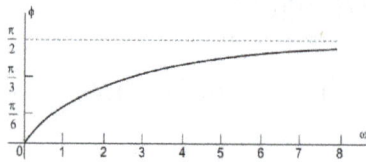

Figure 2

The magnitude $|G(i\omega)|$ of the frequency response can be interpreted geometrically by observing from Eq. (24) that the magnitude of the force in the spring is

$$|F_s(t)| = k|x(t)| = kA|G(i\omega)|$$

28

Moreover, from Eqs. (15), the magnitude of the harmonic excitation is

$$|F(t)| = kA$$

29

Hence, combining Eqs. (28) n and (29) , we can write

$$|G(i\omega)| = \frac{|F_s(t)|}{|F(t)|}$$

30

Or, the magnitude of the frequency response is equal to ratio of the magnitude of the spring force $F_s(t)$ to the magnitude of the excitation force $F(t)$.

Response Due to Harmonic Excitation Forces

$$|F(t)| = F_0 \sin\omega t$$

$\uparrow$

amplitude of forces

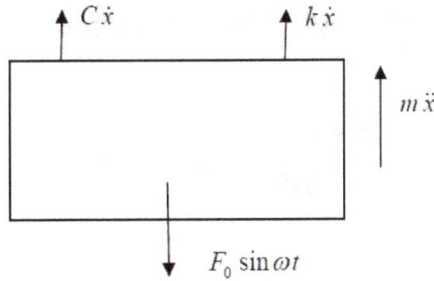

Equilibrium Equation

$$m\ddot{x} + C\dot{x} + kx = F_0 \sin \omega t$$

Homogeneous solution → transient solution → decays exponentially with time

$$m\ddot{x} + C\dot{x} + kx = 0$$

$$\Rightarrow x = A \exp\left(-\xi \omega_n t\right) \sin\left(\omega_d t + \phi\right)$$

On the other hand particular solution does not vanish as time unfolds and is known as steady solution.

Particular solution →A steady state oscillation of frequency ω. We can assume the particular solution to be of the form

$$x = X \sin\left(\omega t - \phi\right)$$

Where ϕ is the angle by which the displacement x lags the force F_0 or the response lags behind the applied force.

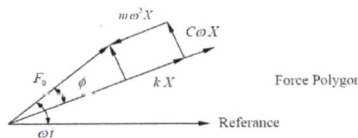

Force Polygon

$$x = X \sin\left(\omega t - \phi\right)$$

$$\dot{x} = \omega X \cos\left(\omega t - \phi\right) = \omega X \sin\left(\omega t - \phi + \frac{\pi}{2}\right)$$

$$\ddot{x} = \omega^2 X \sin\left(\omega t - \phi + \pi\right)$$

Considering the above vectorial representation we can write

$$\left(kX - m\omega^2 X\right)^2 + \left(C\omega X\right)^2 = F_0^2$$

$$X = \frac{F_0}{\sqrt{\left(k - m\omega^2\right)^2 + \left(C\omega\right)^2}}$$

$$\tan\phi = \frac{C\omega}{k - m\omega^2}$$

Complete solution

$$x = \underbrace{A\exp\left(-\xi\,\omega_n t\right)\sin\left(\omega_d t + \phi\right)}_{\text{Transient Solution}} + \underbrace{\frac{F_0\sin\left(\omega t - \phi\right)}{\left[\left(k - m\omega^2\right)^2 + \left(C\omega\right)^2\right]^{\frac{1}{2}}}}_{\text{Steady state oscillation}}$$

The total response is the sum of

I) Starting transient (complementary function) response which decreases exponentially with time and

II) Steady state response (particular integral)

If the first few cycles are of interest, the total response must be studied. But in most cases, the response after starting transient has died away, is of interest.

Steady State Response

Static Response $\rightarrow A_{st} = \dfrac{F_0}{k}$

$\therefore \dfrac{A}{A_{st}}$, Dynamic amplification factor $= \dfrac{1}{\left[\left\{1 - \left(\dfrac{\omega}{\omega_n}\right)^2\right\}^2 + \left(\dfrac{C\omega}{k}\right)^2\right]^{\frac{1}{2}}}$

Notifing that $\omega_n = \sqrt{\dfrac{k}{m}}$ $C_c = \dfrac{2k}{\omega_n}$ $\xi = \dfrac{C}{C_c}$

$$r = Frequency\,ratio\,\frac{\omega}{\omega_n}\ and\ \frac{C\omega}{k} = \frac{C_c\,\xi\,\omega}{k} = \frac{2k}{\omega_n}\cdot\frac{\xi\,\omega}{k} = 2\xi r$$

After substituting the above relations, the dynamic amplification factorbecomes

$$\frac{A}{A_{st}} = \frac{1}{\left\{\left(1-r^2\right)+\left(2\xi r\right)^2\right\}^{1/2}}$$

Dynamic magnification factor in terms of r and ξ

$$\tan\phi = \frac{2\xi r}{1-r^2}$$

Response of Second Order Systems to Harmonic Excitation (Algebraic Approach for Particular Integral)

$$m\ddot{x}(t) + C\dot{x}(t) + kx(t) = F(t)$$

$$F(t) = kf(t) = kA\sin\omega t$$

Where ω is the excitation frequency, sometimes referred to as driving frequency. The above equation can be rewritten as

$$m\ddot{x}(t) + C\dot{x}(t) + kx(t) = kA\sin\omega t$$

or, $\ddot{x}(t) + 2\xi\omega_n\dot{x}(t) + \omega_n^2 x(t) = A\omega_n^2\sin\omega t$

Let the solution is $x(t) = C_1\sin\omega t + C_2\cos\omega t$

Substituting this in the above equation we can write the left hand side (LHS) as

$$-\omega^2\left(C_1\sin\omega t + C_2\cos\omega t\right) + 2\zeta\omega\omega_n\left(C_1\cos\omega t - C_2\sin\omega t\right) + \omega_n^2\left(C_1\sin\omega t + C_2\cos\omega t\right)$$

$$\Rightarrow \left(\omega_n^2 - \omega^2\right)\left(C_1\sin\omega t + C_2\cos\omega t\right) + 2\zeta\omega\omega_n\left(C_1\cos\omega t - C_2\sin\omega t\right)$$

$$\Rightarrow \left\{C_1\left(\omega_n^2 - \omega^2\right) - C_2 2\zeta\omega\omega_n\right\}\sin\omega t + \left\{C_2\left(\omega_n^2 - \omega^2\right) + C_1 2\zeta\omega\omega_n\right\}\cos\omega t$$

Equating the coefficient of $\sin\omega t$ & $\cos\omega t$ respectively on the both sides of the equation, we obtain two algebraic equations

$$\left(\omega_n^2 - \omega^2\right)C_1 - 2\zeta\omega\omega_n C_2 = \omega_n^2 A \rightarrow(1)$$

$$2\zeta\omega\omega_n C_1 + \left(\omega_n^2 - \omega^2\right)C_2 = 0 \rightarrow(2)$$

Solving the above equations

$$(2) \Rightarrow C_1 = -\frac{\omega_n^2 - \omega^2}{2\zeta\omega\omega_n} C_2$$

Substituting in equation (1) we get

$$\left(\omega_n^2 - \omega^2\right) \times \left(-\frac{\omega_n^2 - \omega^2}{2\zeta\omega\omega_n}\right) C_2 - 2\zeta\omega\omega_n C_2 = \omega_n^2 A$$

$$\Rightarrow -C_2 \left\{\frac{\left(\omega_n^2 - \omega^2\right)^2 + 4\zeta^2\omega^2\omega_n^2}{2\zeta\omega\omega_n}\right\} = \omega_n^2 A$$

$$\Rightarrow C_2 = -\frac{2\zeta\omega\omega_n}{\left(\omega_n^2 - \omega^2\right)^2 + \left(2\zeta\omega\omega_n\right)^2} = \omega_n^2 A = \frac{2\zeta\left(\dfrac{\omega}{\omega_n}\right)}{\left\{1 - \left(\dfrac{\omega}{\omega_n}\right)^2\right\}^2 + \left(2\zeta\dfrac{\omega}{\omega_n}\right)^2} A$$

$$\text{and } C_1 = -\frac{\omega_n^2 - \omega^2}{2\zeta\omega\omega_n} \times \left\{-\frac{2\zeta\omega\omega_n}{\left(\omega_n^2 - \omega^2\right)^2 + \left(2\zeta\omega\omega_n\right)^2}\right\} \omega_n^2 A$$

$$= \frac{\omega_n^2 - \omega^2}{\left(\omega_n^2 - \omega^2\right)^2 + \left(2\zeta\omega\omega_n^2\right)} \omega_n^2 A = \frac{1 - \left(\dfrac{\omega}{\omega_n}\right)^2}{\left\{1 - \left(\dfrac{\omega}{\omega_n}\right)^2\right\}^2 + \left(2\zeta\dfrac{\omega}{\omega_n}\right)^2}$$

$$\therefore x(t) = \frac{A}{\left\{1 - \left(\dfrac{\omega}{\omega_n}\right)^2\right\}^2 + \left(2\zeta\dfrac{\omega}{\omega_n}\right)^2} \left[\left\{1 - \left(\dfrac{\omega}{\omega_n}\right)^2\right\} \sin\omega t - 2\zeta\left(\dfrac{\omega}{\omega_n}\right)\cos\omega t\right]$$

Define

$$\frac{1 - \left(\dfrac{\omega}{\omega_n}\right)^2}{\left[\left\{1 - \left(\dfrac{\omega}{\omega_n}\right)^2\right\}^2 + \left(2\zeta\dfrac{\omega}{\omega_n}\right)^2\right]^{\frac{1}{2}}} = \cos\phi$$

$$\frac{2\zeta\dfrac{\omega}{\omega_n}}{\left[\left\{1-\left(\dfrac{\omega}{\omega_n}\right)^2\right\}^2+\left(2\zeta\dfrac{\omega}{\omega_n}\right)^2\right]^{\frac{1}{2}}}=\sin\phi$$

$$\therefore x(t)=X(\omega)\left(\sin\omega t\cos\phi-\sin\varphi\cos\omega t\right)$$

$$=X(\omega)\sin(\omega t-\phi)$$

$$\text{Where, }=X(\omega)=\frac{A}{\left[\left\{1-\left(\dfrac{\omega}{\omega_n}\right)^2\right\}^2+\left(2\zeta\dfrac{\omega}{\omega_n}\right)^2\right]^{\frac{1}{2}}}$$

$$\text{and }\phi=\tan^{-1}\frac{2\zeta\dfrac{\omega}{\omega_n}}{1-\left(\dfrac{\omega}{\omega_n}\right)^2}$$

Next, let us reproduce the above results by working with complex vectors. Hence, we consider

$$m\ddot{x}(t)+c\dot{x}(t)+kx=F_0e^{i\omega t}$$

Then, letting the steady state response have the form

$$x(t)=X_0e^{i(\omega t-\phi)}=X_0e^{-i\phi}e^{i\omega t}=\bar{X}_0e^{i\omega t}$$

Substituting the above equation we can write

$$\left[-m\omega^2+k+i\omega C\right]\bar{X}_0=F_0$$

$$\Rightarrow \bar{X}_0=\frac{F_0}{\left(k-m\omega^2\right)+i\omega C}$$

$$=\frac{F_0/k}{\left(1-\dfrac{m}{k}.\omega^2\right)+i\omega\dfrac{C}{k}}\quad\text{where }F_0/k=X_{st}$$

$$= \frac{F_0/k}{\left\{1-\left(\omega/\omega_n\right)^2\right\}+i\cdot 2\xi\left(\omega/\omega_n\right)}$$

$$\frac{\bar{X}_0}{X_{st}} = \frac{1}{\left\{1-\left(\omega/\omega_n\right)^2\right\}+i\cdot 2\xi\left(\omega/\omega_n\right)}$$

$$\Rightarrow \left|\frac{\bar{X}_0}{X_{st}}\right| = \frac{1}{\left[\left\{1-\left(\omega/\omega_n\right)^2\right\}^2+\left\{2\xi\left(\omega/\omega_n\right)\right\}^2\right]^{1/2}} = \frac{1}{\left[\left(1-r^2\right)^2+\left(2\xi r^2\right)^2\right]^{1/2}}$$

$$H(\omega) = \left|\frac{\bar{X}_0}{X_{st}}\right|$$

$$\tan\phi = \frac{2\xi\,\omega/\omega_n}{1-\left(\omega/\omega_n\right)^2} = \frac{2\xi r}{1-r^2}$$

where, $\left|\dfrac{\bar{X}_0}{X_{st}}\right|$ and ϕ are magnifucation factor and phase angle respectively

To find the value at which the peaks of the curve occur, we use the standard technique of calculus, i.e., we differentiate the equation of magnification factor with respect to ω and set the result equal to zero.

$$\text{Let } M = \left[1-\left(\frac{\omega}{\omega_n}\right)^2\right]^2+\left(2\xi\,\omega/\omega_n\right)^2$$

$$\frac{\partial M}{\partial\omega} = 2\left(0-\frac{\omega}{\omega_n^2}\right)\left[1-\left(\frac{\omega}{\omega_n}\right)^2\right]+4\frac{\xi}{\omega_n}\times\left(2\xi\frac{\omega}{\omega_n}\right)=0$$

$$\Rightarrow -\frac{\omega}{\omega_n^2}\left[1-\left(\frac{\omega}{\omega_n}\right)^2\right]+2\xi^2\frac{\omega}{\omega_n^2}=0$$

$$\Rightarrow -\frac{\omega}{\omega_n^2}\left[\left(1-2\xi^2\right)-\left(\frac{\omega}{\omega_n}\right)^2\right]=0$$

$$\Rightarrow either \quad \frac{\omega}{\omega_n^{\,2}} = 0 \quad or \quad \left(1 - 2\xi^2\right) - \left(\frac{\omega}{\omega_n}\right)^2 = 0$$

$$\frac{\omega}{\omega_n^{\,2}} \neq 0 \implies \frac{\omega}{\omega_n} = \sqrt{1 - 2\xi^2}$$

$$or, \quad \omega = \omega_n \sqrt{1 - 2\xi^2}$$

Indicating that the maxima do not occur at the undamped natural requency ω_n but for $\omega/\omega_n < 1$ depending on the amount of damping.

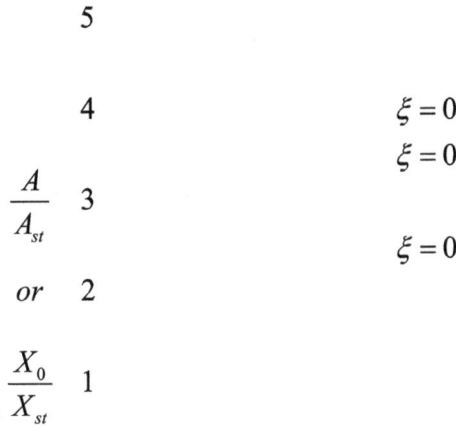

5

4 $\xi = 0$
                                                     $\xi = 0$
$\dfrac{A}{A_{st}}$ 3
                                                     $\xi = 0$
or 2

$\dfrac{X_0}{X_{st}}$ 1

0.5   1.0   1.5   2.0   2.5   3.0

$$r = \omega/\omega_n$$

$$\left|\frac{X}{X_{st}}\right| = \frac{1}{\left[\left\{1 - \left(\frac{\omega}{\omega_n}\right)^2\right\}^2 + \left(2\xi\frac{\omega}{\omega_n}\right)^2\right]^{1/2}}$$

$$\left|\frac{X}{X_{st}}\right|_{max} \quad at \quad \frac{\omega}{\omega_n} = \sqrt{1 - 2\xi^2}$$

$$\left|\frac{X}{X_{st}}\right|_{max} = \frac{1}{\left\{\left(1 - 1 + 2\xi^2\right)^2 + 4\xi^2\left(1 - 2\xi^2\right)\right\}^{1/2}}$$

$$= \frac{1}{\left\{4\xi^4 + 4\xi^2 - 8\xi^4\right\}^{1/2}}$$

$$\approx \frac{1}{\left\{4\xi^2\left(1-\xi^2\right)\right\}^{1/2}}$$

$$\approx \frac{1}{2\xi\left(1-\xi^2\right)^{1/2}}$$

$$\approx \frac{1}{2\xi}$$

We notice that for light damping, such as when $\xi < 0.05$, the maximum of $\left|\dfrac{X}{X_{st}}\right|$ occurs in the immediate neighborhood of $\omega / \omega_n = 1$ introducing the notation $\left|\dfrac{X}{X_{st}}\right|_{max} = Q$ we obtain for small values of ξ

$$Q \cong \frac{1}{2\xi}$$

The symbol Q is known as the quality factor because many electrical applications, such as the tuning circuit of a radio, the interest lies in an amplitude at resonance that is as large as possible the symbol is often referred to as the 'Q' of the circuit.

Frequency Response

Frequency response is the quantitative measure of the output spectrum of a system or device in response to a stimulus, and is used to characterize the dynamics of the system. It is a measure of magnitude and phase of the output as a function of frequency, in comparison to the input. In simplest terms, if a sine wave is injected into a system at a given frequency, a linear system will respond at that same frequency with a certain magnitude and a certain phase angle relative to the input. Also for a linear system, doubling the amplitude of the input will double the amplitude of the output. In addition, if the system is time-invariant (so LTI), then the frequency response also will not vary with time. Thus for LTI systems, the frequency response can be seen as applying the system's transfer function to a purely imaginary number argument representing the frequency of the sinusoidal excitation.

Two applications of frequency response analysis are related but have different objectives. For an audio system, the objective may be to reproduce the input signal with no distortion. That would require a uniform (flat) magnitude of response up to the bandwidth limitation of the system, with the signal delayed by precisely the same amount of time at all frequencies. That amount of time could be seconds, or weeks or months in the case of recorded media. In contrast, for a feedback apparatus used to control a dynamic system, the objective is to give the closed-loop system improved response as compared to the uncompensated system. The feedback generally needs to respond to system dynamics within a very small number of cycles of oscillation (usually less than one full cycle), and with a definite phase angle relative to the commanded control input. For feedback of sufficient amplification, getting the phase angle wrong can lead to instability for an open-loop stable system, or failure to stabilize a system that is open-loop unstable. Digital filters may be used for both audio systems and feedback control systems, but since the objectives are different, generally the phase characteristics of the filters will be significantly different for the two applications.

Estimation and Plotting

Estimating the frequency response for a physical system generally involves exciting the system with an input signal, measuring both input and output time histories, and comparing the two through a process such as the Fast Fourier Transform (FFT). One thing to keep in mind for the analysis is that the frequency content of the input signal must cover the frequency range of interest or the results will not be valid for the portion of the frequency range not covered.

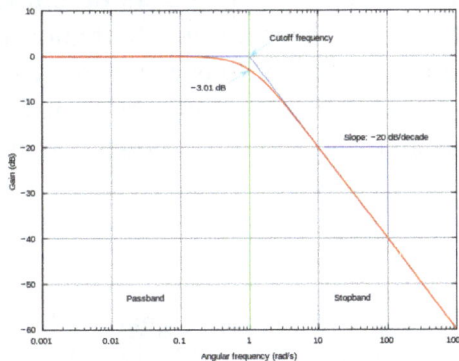

Frequency response of a low pass filter with 6 dB per octave or 20 dB per decade

The frequency response of a system can be measured by applying a *test signal*, for example:

- applying an impulse to the system and measuring its response

- sweeping a constant-amplitude pure tone through the bandwidth of interest and measuring the output level and phase shift relative to the input

- applying a signal with a wide frequency spectrum (for example digitally-generated maximum length sequence noise, or analog filtered white noise equivalent, like pink noise), and calculating the impulse response by deconvolution of this input signal and the output signal of the system.

The frequency response is characterized by the *magnitude* of the system's response, typically measured in decibels (dB) or as a decimal, and the *phase*, measured in radians or degrees, versus frequency in radians/sec or Hertz (Hz).

These response measurements can be plotted in three ways: by plotting the magnitude and phase measurements on two rectangular plots as functions of frequency to obtain a Bode plot; by plotting the magnitude and phase angle on a single polar plot with frequency as a parameter to obtain a Nyquist plot; or by plotting magnitude and phase on a single rectangular plot with frequency as a parameter to obtain a Nichols plot.

For audio systems with nearly uniform time delay at all frequencies, the magnitude versus frequency portion of the Bode plot may be all that is of interest. For design of control systems, any of the three types of plots [Bode, Nyquist, Nichols] can be used to infer closed-loop stability and stability margins (gain and phase margins) from the open-loop frequency response, provided that for the Bode analysis the phase-versus-frequency plot is included.

Nonlinear Frequency Response

If the system under investigation is nonlinear then applying purely linear frequency domain analysis will not reveal all the nonlinear characteristics. To overcome these limitations, generalized frequency response functions and nonlinear output frequency response functions have been defined that allow the user to analyze complex nonlinear dynamic effects. The nonlinear frequency response methods reveal complex resonance, inter modulation, and energy transfer effects that cannot be seen using a purely linear analysis and are becoming increasingly important in a nonlinear world.

Applications

In electronics this stimulus would be an input signal. In the audible range it is usually referred to in connection with electronic amplifiers, microphones and loudspeakers. Radio spectrum frequency response can refer to measurements of coaxial cable, twisted-pair cable, video switching equipment, wireless communications devices, and antenna systems. Infrasonic frequency response measurements include earthquakes and electroencephalography (brain waves).

Frequency response requirements differ depending on the application. In high fidelity audio, an amplifier requires a frequency response of at least 20–20,000 Hz, with a tolerance as tight as ±0.1 dB in the mid-range frequencies around 1000 Hz, however, in telephony, a frequency response of 400–4,000 Hz, with a tolerance of ±1 dB is sufficient for intelligibility of speech.

Frequency response curves are often used to indicate the accuracy of electronic components or systems. When a system or component reproduces all desired input signals with no emphasis or attenuation of a particular frequency band, the system or component is said to be "flat", or to have a flat frequency response curve.

Once a frequency response has been measured (e.g., as an impulse response), provided the system is linear and time-invariant, its characteristic can be approximated with arbitrary accuracy by a digital filter. Similarly, if a system is demonstrated to have a poor frequency response, a digital or analog filter can be applied to the signals prior to their reproduction to compensate for these deficiencies.

$$\left|\frac{X}{X_{st}}\right| = \frac{1}{\left[\left\{1 - \left(\frac{\omega}{\omega_n}\right)^2\right\}^2 + \left(2\xi\frac{\omega}{\omega_n}\right)^2\right]^{1/2}}$$

$$\tan\phi = \frac{2\xi\,\omega/\omega_n}{1 - \left(\omega/\omega_n\right)^2}$$

These indicate that non-dimensional amplitude $\left|\dfrac{X}{X_{st}}\right|$ and the phase angle ϕ are functions of the frequency ratio n $\dfrac{\omega}{\omega_n}$ and the damping factor ξ.

There are three cases possible:

(i) $\dfrac{\omega}{\omega_n} \ll 1$ The inertia and damping forces are small, which results in a small phase angle ϕ. The magnitude of the impressed force is then nearly equal to the spring force. This zone is called as stiffness dominated zone.

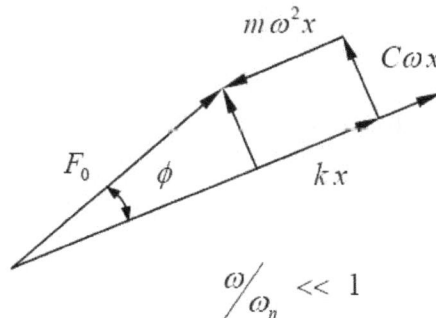

(ii) For $\omega/\omega_n = 1.0$, the phase angle is $90°$ and the force diagram appears as shown in the figure.

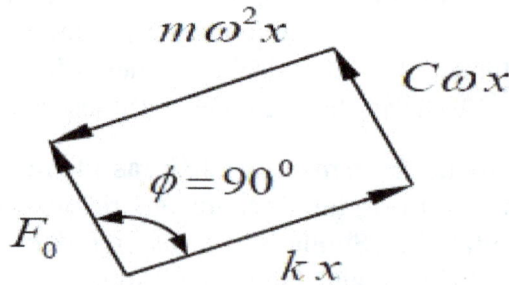

$$\omega/\omega_n = 1$$

This is called as damping dominated zone.

The inertia force is now larger, is balanced by the spring force, whereas the impressed force overcomes the damping force. The amplitude at resonance can be found to be

$$X = \frac{F_0}{C\omega} = \frac{F_0}{2\xi k}$$

(iii) At large value of $\omega/\omega_n \gg 1$, ϕ approaches $180°$, and the impressed force expanded almost entirely in overcoming the large inertia force as shown in Figure:

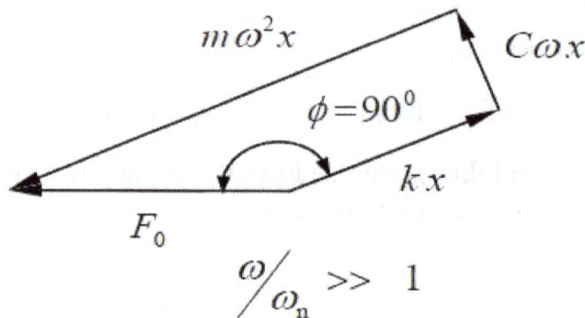

$$\omega/\omega_n \gg 1$$

This is called as inertia dominated zone.

Summary of some Relations

1. $\dfrac{d}{dr}\left(\dfrac{A}{A_{st}}\right) = 0$

$$\Rightarrow r^2 = 1 - 2\xi^2$$

For practical values of damping, that is small damping, magnification factor is maximum, when r is slightly less than unity.

When $\xi = 0 \rightarrow$

2. When $r = 1$, $\dfrac{\overline{A}}{A_{st}} = \dfrac{1}{2\xi_i}$

3. $\overline{A}$ is very small when $\omega \gg \omega_n$

4. When $\overline{A} \rightarrow A_{st}$ $\omega \ll \omega_n$

5. (i) $\phi = 0$ at $\xi = 0$ and $\omega \gg \omega_n$

    i.e. the response is in phase with exciting force.

    (ii) $\phi = \dfrac{\pi}{2}$ at the resonant condition i.e. r = 1

    (iii) $\phi = \pi$, when $\omega \gg \omega_n$

    the response is out of phase with the exciting frequency

Rotating Unbalance

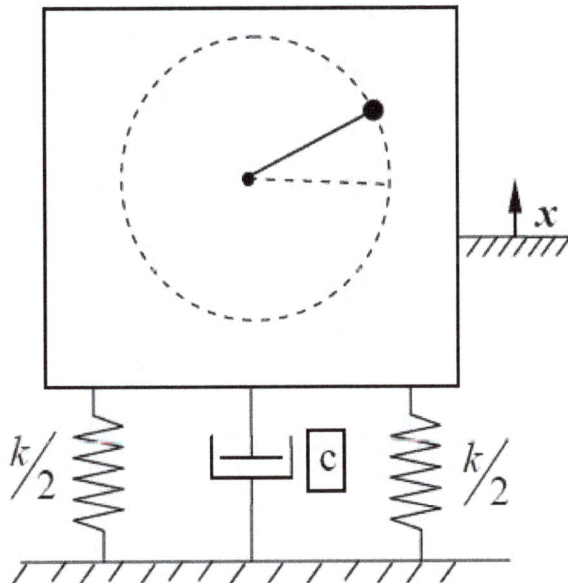

Total mass of the system = M

Rotating mass = m

Effective mass of non rotating system = M − m

The displacement of rotating mass $= x + e \sin \omega t$

The equation of motion is then

$$(M - m)\ddot{x} + m\frac{d^2}{dt^2}(x + e\sin \omega t) = -kx - c\dot{x}$$

Which can be rearranged as

$$M\ddot{x} + kx + c\dot{x} = \left(me\,\omega^2\right)\sin \omega t$$

When $x = X\sin(\omega t - \phi)$

$$X = \frac{me\,\omega^2}{\sqrt{\left(k - \omega^2 M\right)^2 + (c\omega)^2}} \quad \Rightarrow \quad \frac{M}{m}\cdot\frac{X}{e} = \frac{\left(\omega/\omega_n\right)^2}{\sqrt{\left\{1 - \left(\dfrac{\omega}{\omega_n}\right)^2\right\}^2 + \left(2\zeta\dfrac{\omega}{\omega_n}\right)^2}}$$

$$\tan\phi = \frac{C\omega}{k - \omega^2 M} = \frac{2\zeta\left(\omega/\omega_n\right)}{1 - \left(\omega/\omega_n\right)^2}$$

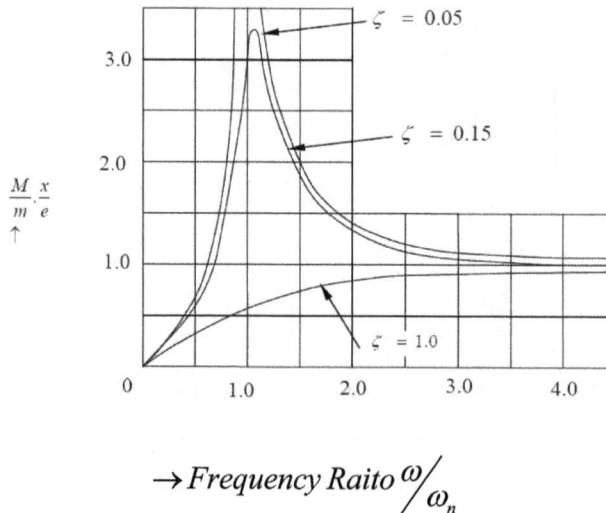

$$\rightarrow Frequency\ Raito\ \omega/\omega_n$$

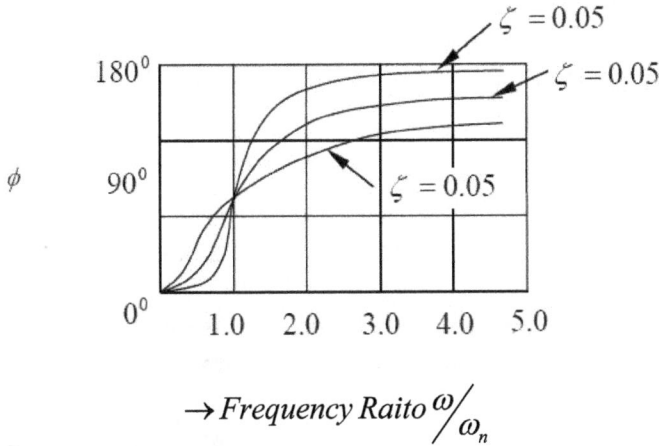

$$\rightarrow Frequency\ Raito\ \omega\!\!\Big/\!\!\omega_n$$

The complete solution is given by

$$x(t) = X_1 e^{-\zeta\omega_n t} \sin\left(\sqrt{1-\zeta^2}\,\omega_n t + \phi\right)$$

$$+\frac{me\,\omega^2}{\sqrt{\left(k-M\,\omega^2\right)^2+\left(C\omega\right)^2}}\sin\left(\omega t - \phi\right)$$

Example

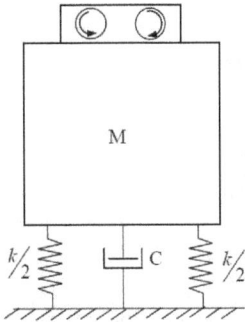

Resonant amplitude = 0.60 cm

The amplitude appeared to approach a fixed value of 0.08cm

beyond resonant peak.

The resonant amplitude is

$$X - \frac{me\!\!\Big/\!\!M}{2\zeta} - 0.60$$

When $\omega \quad \omega$ the response amplitude becomes

$$X = \frac{me}{M} = 0.08\,cm$$

Solving the two equations simultaneously, the damping factor of the system is

$$\zeta = \frac{0.08}{2 \times 0.6} = 0.0666$$

Example

A tail rotor of a helicopter is supported on a tail section having stiffness, $k = 1 \times 10^5\ N/m$ in the vertical direction and effective mass of 20 kg. Calculate the deflection of the tail section when the tail rotor rotates at 1500 rpm with an eccentric mass of 0.5 kg attached to a blade of arm length 15 cm.

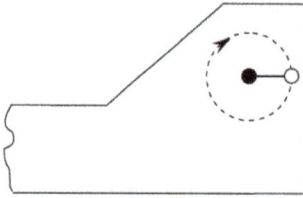

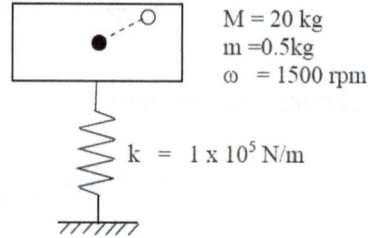

M = 20 kg
m = 0.5kg
ω = 1500 rpm

k = 1 x 10⁵ N/m

$$X = \frac{m e\, \omega^2}{\sqrt{\left(k - \omega^2\, M\right)^2 + \left(C\omega\right)^2}}$$

$$C = 0$$

$$= \frac{m e\, \omega^2 / k}{1 - \left(\dfrac{\omega}{\omega_n}\right)^2}$$

$$e = 15\,cm = 0.15\,m$$

$$\omega = \frac{1500 \times 2\pi}{60} = 157.08\,rad/s$$

$$= \frac{0.15 \times 0.5 \times 157.08^2}{(1 - 4.9348) \times 10^5}$$

$$\omega_n = \sqrt{\frac{k}{m}} = \sqrt{\frac{1 \times 10^5}{20}} = 70.7\,rad/s$$

$$= 4.703 \times 10^{-3}\, m$$

$$\left(\frac{\omega}{\omega_n}\right)^2 = \left(\frac{157.08}{70.7}\right)^2 = 4.9348$$

Exercise 1:

Plot the normalized amplitude $\left(\dfrac{A}{A_{st}}\right)$ and phase ϕ of the steady state response of a

damped system vs frequency ratio $r = \dfrac{\omega}{\omega_n}$ for the force vibration of the system of given parameters.

$$\omega_n = 10\ rad/s$$

$$\omega = 5\ rad/s$$

$$\xi = 0.01, 0.1, 0.5, 1.0$$

$$F_0 = 100\,N,\ m = 100\,kg$$

Initial condition $x(0) = 0.05,\ X(0) = 0$

$$\omega_n = \sqrt{\frac{k}{m}} \Rightarrow k = 10^2 \times 100 = 10000\,N/m$$

$$r = \frac{\omega}{\omega_n} = \frac{1}{2}$$

Support Motion

There are several practical problems that can be idealized to calculate response due to support motion, such as building structures subjected to earth quake or systems

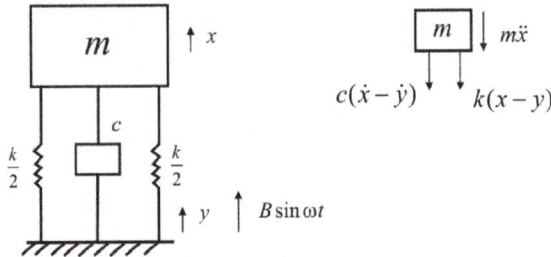

$$m\ddot{x} = -k(x-y) - c(\dot{x}-\dot{y})$$

$$\Rightarrow m\ddot{x} + c\dot{x} + kx = ky + c\dot{y}$$

Let $y = B\sin\omega t\ \&\ x = A\sin(\omega t - \varphi)$

$$\Rightarrow m\ddot{x} + c\dot{x} + kx = kB\sin\omega t + c\omega B\sin\left(\omega t + \frac{\pi}{2}\right)$$

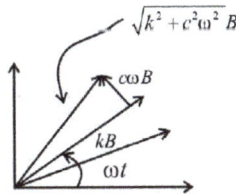

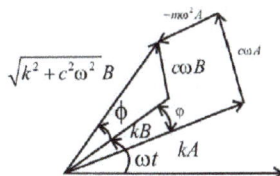

$$\overline{k^2 + c^2\omega^2}\, B = \sqrt{\left(k - m\omega^2\right)^2 + c^2\,\omega^2}.A$$

$$\frac{A}{B} = \frac{\sqrt{k^2 + c^2\omega^2}}{\sqrt{\left(k - m\omega^2\right)^2 + c^2\,\omega^2}}$$

$$= \sqrt{\frac{1 + \left(2\xi r\right)^2}{\left(1 - r^2\right)^2 + \left(2\xi r\right)^2}}$$

$\Rightarrow$ transmissibility of motion

$$\tan\left(\phi + \varphi\right) = \frac{c\omega A}{KA - m\omega^2 A}$$

$$= \frac{c\omega}{K - m\omega^2}$$

$$\tan\phi = \frac{c\omega B}{K B} = \frac{c\omega}{K}$$

$$\tan\left(\phi + \varphi\right) = \frac{\tan\phi\ \tan\varphi}{1 - \tan\phi\tan\varphi} = \frac{\omega}{K - m\omega}$$

$$\tan\varphi = \frac{\dfrac{m}{k}c\omega^3}{k - m\omega^2 + \dfrac{c^2\omega^2}{k}}$$

$$= \frac{mc\omega^3}{k\left(k - m\omega^2\right) + c^2\omega^2}$$

$$c\omega\left(1 - \tan\varphi.\frac{c\omega}{k}\right) = \left(k - m\omega^2\right)\left(\frac{c\omega}{k} + \tan\varphi\right)$$

$$\Rightarrow \left(k - m\omega^2 + \frac{c^2\omega^2}{k}\right)\tan\varphi = \frac{m}{k}c\omega^3$$

$$\tan \varphi = \frac{\dfrac{m}{k} c\omega^3}{k - m\omega^2 + \dfrac{c^2 \omega^2}{k}}$$

$$= \frac{\dfrac{m}{k} \cdot \dfrac{c}{k} \omega^3}{1 - \dfrac{m}{k}\omega^2 + \left(\dfrac{c^2}{k^2}\right)\omega^2}$$

$$= \frac{2\zeta \left(\dfrac{\omega}{\omega_n}\right)^3}{1 - \left(\dfrac{\omega}{\omega_n}\right)^2 + \left(2\zeta \dfrac{\omega}{\omega_n}\right)^2}$$

$$As \quad \frac{c}{k} = \frac{\zeta c_c}{k} = \frac{2\zeta}{\omega_n} \qquad \frac{m}{k} = \frac{1}{\omega_n^2}$$

$$c_c = 2m\omega_n = \frac{2k}{\omega_n}$$

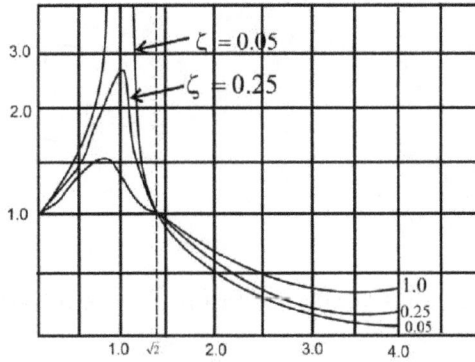

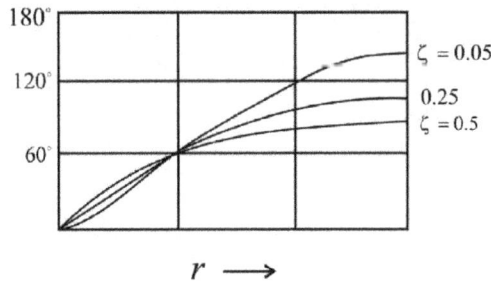

$$r \longrightarrow$$

Vibration Isolation

The situation of vibration system is just reverse of base isolation system. In this system is vibration and our interest is to find out the vibration transmitted to the base.

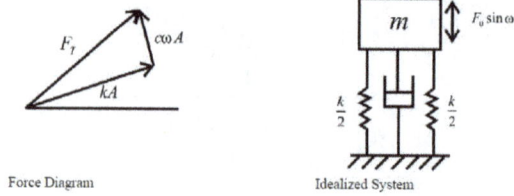

Force Diagram

Idealized System

Transmitted force $F_T = \sqrt{(kA)^2 + (c\omega A)^2}$

$$= kA\sqrt{1 + \left(\frac{c\omega}{k}\right)^2} = kA\sqrt{1 + (2\varsigma r)^2}$$

When the disturbing force equal to $F_0 \sin \omega t$, the value of A, amplitude in the preceding equation is

$$A = \frac{F_0/k}{\sqrt{(1-r^2)^2 + (2\varsigma r)^2}}$$

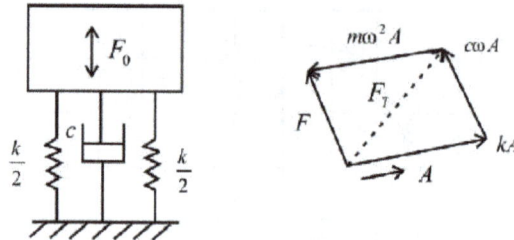

The transmissibility TR, defined as the ratio of the transmitted force to that of the disturbing force, is then

$$TR = \left|\frac{F_T}{F_0}\right| = \sqrt{\frac{1 + (2\varsigma r)^2}{(1-r^2) + (2\varsigma r)^2}}$$

Comparing the preceding equation shows that

$$TR = \left|\frac{F_T}{F_0}\right| = \left|\frac{A}{B}\right|$$

When damping is negligible, the transmissibility reduces to

$$TR = \frac{1}{r^2 - 1}$$

Here it is understood that the value of $r\left(\dfrac{\omega}{\omega_n}\right)$ to be used is always greater than $\sqrt{2}$

A machine of 100 kg mass is supported on springs of total stiffness of 700b KN/m and has an unbalanced rotating element, which results on a disturbing force of 350 N at a speed of 3000 rev/min. Assuming $\xi = 0.20$ determine (i) its ampl. of motion due to the unbalance (ii) the transmissibility and (iii) the transmitted force.

Solution: The statical deflection of the system is

$$\frac{100 \times 9.81}{700 \times 10^3} = 1.401 \times 10^{-3} = 1.401\,mm$$

and its natural frequency is found to be

$$f_n = \frac{1}{2\pi}\sqrt{\frac{9.81}{1.401 \times 10^{-3}}} = 13.32\,Hz \qquad \begin{bmatrix} f_n = \dfrac{1}{2\pi}\sqrt{\dfrac{k}{m}} \\[2mm] k\Delta = mg \Rightarrow \dfrac{k}{m} = \dfrac{g}{\Delta} \end{bmatrix}$$

The amplitude of vibration is

$$a = \frac{F_0 / k}{\sqrt{\left(1-r^2\right)^2 + \left(2\zeta r\right)^2}} = \frac{350/700 \times 10^3}{\sqrt{\left(1-\left(\dfrac{50}{13.32}\right)^2\right)^2 + \left(2 \times 0.20 \times \dfrac{50}{13.32}\right)^2}}$$

$$= 3.79 \times 10^{-5}\,m$$

$$= 0.0379\,mm$$

The transmissibility

$$TR = \frac{\sqrt{1+\left(2\zeta r\right)^2}}{\sqrt{\left(1-r^2\right)+\left(2\zeta r\right)^2}} = \frac{\sqrt{1+\left(2 \times 0.20 \times \dfrac{50}{13.32}\right)^2}}{\left[\left\{1-\left(\dfrac{50}{13.32}\right)^2\right\}+\left\{2 \times 0.20 \times \dfrac{50}{13.32}\right\}^2\right]^{1/2}} = 0.137$$

The transmitted force, $F_{TR} = 350 \times 0.137 = 47.89\,N$

Vibration Measuring Instrument

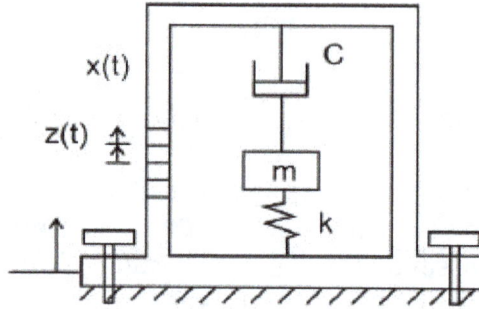

To determine the behavior of such instruments we consider the equation of motion of m, which is $m\ddot{x} = -c(\dot{x}-\dot{y})-k(x-y)$

Here x and y are the displacement of the seismic mass and the vibrating body, respectively, both measured w.r.t an inertial reference. Let the relative displacement

$$z = (x-y)$$

And assuming sinusoidal motion $y = Y\sin\omega t$ of the vibrating base, we obtain the equation

$$m\ddot{z} + c\dot{z} + kz = m\omega^2 Y \sin\omega t$$

The steady state solution $z = Z\sin(\omega t - \phi)$ is then available as

$$z = \frac{m\omega^2 Y}{\left[\left(k-m^2\right)^2+(c\omega)^2\right]^{\frac{1}{2}}} = \frac{Y\left(\dfrac{\omega}{\omega_n}\right)^2}{\left[\left\{1-\left(\dfrac{\omega}{\omega_n}\right)^2\right\}^2+\left(2\zeta\dfrac{\omega}{\omega_n}\right)^2\right]^{\frac{1}{2}}}$$

$$\text{and } \tan\phi = \frac{\omega c}{k-m\omega^2} = \frac{2\zeta\dfrac{\omega}{\omega_n}}{1-\left(\dfrac{\omega}{\omega_n}\right)^2}$$

It is evident then that the parameters involved are the frequency ratio $\left(\dfrac{\omega}{\omega_n}\right)$ and the

damping factor ξ. Depending on the natural frequency we can have two different instrument.

1. Seismometer - instrument with low natural frequency for

$$\frac{\omega}{\omega_n} >> 0 \left[\left\{ 1 - \left(\frac{\omega}{\omega_n} \right)^2 \right\}^2 \right]^{\frac{1}{2}} \approx \left(\frac{\omega}{\omega_n} \right)^2$$

$$\Rightarrow \frac{\omega}{\omega_n} = 100, \zeta = 0$$

$$\Rightarrow 100.015$$

$$r = 100 \quad \zeta = 0.0 \rightarrow 100.493 \qquad high$$

For every frequency ratio we can write

$z \approx Y$ regardless of the value of damping

2. Accelerometer – Instrument with high natural frequency

$$\sqrt{ \left(1 - \left(\frac{\omega}{\omega_n} \right)^2 \right)^2 + \left(2\zeta \frac{\omega}{\omega_n} \right)^2 } \rightarrow 1 \, when \, \frac{\omega}{\omega_n} \rightarrow 0$$

$$z = \frac{\omega^2 Y}{\omega_n^2} = \frac{acceleration}{\omega_n^2}$$

Thus z becomes proportional to the acceleration of the motion to be measured with a factor $\frac{1}{\omega_n^2}$.

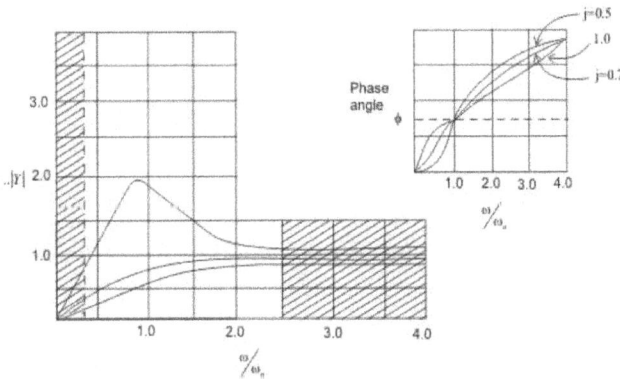

The useful range of range of accelerometer can be seen from magnified plot of

$$\frac{1}{\sqrt{\left(1 - r^2\right)^2 + \left(2\zeta r\right)^2}}$$

a useful frequency range $0 - 20\, Hz$.

Sensitivity charge/g

       volt/g

$$0 < r < 0.2$$
$$Max\, Error = 0.01\%$$

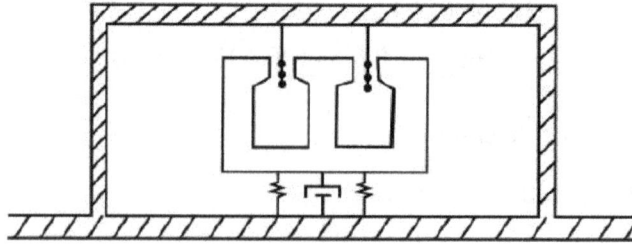

Seismometer

Natural frequency between 1 Hz to 5 Hz and useful frequency range of 10 Hz to 2000 Hz. Sensitivity 20 mv / cm / s to 350 mv/cm/s.

Accelerometer

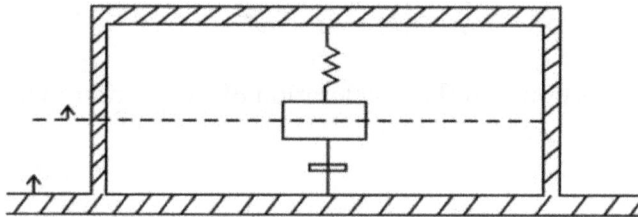

3

Energy Damping in Aircraft Structural Dynamics

Energy damping is a process which occurs in oscillatory systems. The purpose of this process is to discharge energy from the system. The topics discussed in the chapter are of great importance to broaden the existing knowledge on aircraft engineering.

Energy Dissipated by Damping

Usually damping is present in all oscillatory system. Its effect is to remove energy from the system. Energy in a vibrating system is either dissipated into heat sound or radiated away.

In vibration analysis, we are generally concerned with damping in terms of system response. The loss of energy from the oscillatory system results in the decay of amplitude of free vibration. In steady state force vibration, the loss of energy is balanced by the energy which is supplied by the excitation.

A vibrating system may encounter many different types of damping forces, from internal molecular friction to sliding friction and fluid resistance. Generally their mathematical description is quite complicated and not suitable for vibration analysis. Thus simplified damping models have been developed that in many cases are not found to be adequate in evaluating the system response.

Energy dissipated is usually determined under the condition of cyclic oscillation. Depending on the type of damping present, the force-displacement relationship when plotted may differ greatly. In all cases, however, the force-displacement curve will enclose an area, referred to as the hysteresis loop that is proportional to the energy loss per cycle.

The energy lost per cycle due to a damping force F_d is computed from the general

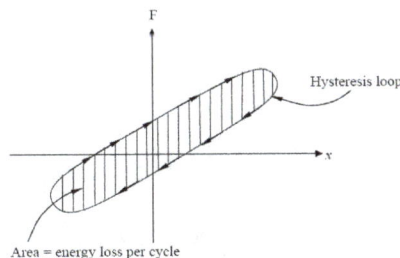

Equation

$$W_d = \oint F_d\, dx \qquad\qquad \dot{X} = \frac{dX}{dt} \Rightarrow dX = \dot{X}dt$$

$$F_d = c\dot{X}$$

$$W_d = \oint c\dot{X}\left(\dot{X}\right)dt$$

$$= \oint c\dot{X}^2\, dt$$

We consider in this section the simplest case of energy dissipation, that of a spring-mass system with viscous damping. The damping force in this case is $F_d = c\dot{X}$. With the steady state displacement and velocity

$$x = X\sin(\omega t - \phi)$$
$$\dot{x} = \omega X\cos(\omega t - \phi)$$
$$\ddot{x} = -\omega^2 X\sin(\omega t - \phi)$$

the energy dissipated per cycle becomes

$$W_d = \oint c\dot{X}\, dx = \oint c\dot{X}^2\, dt$$

$$= c\omega^2 X^2 \int_{0}^{\frac{2\pi}{\omega}} \cos^2(\omega t - \phi)\, dt$$

$$= c\omega^2 X^2 \left(\frac{\pi}{\omega}\right)$$

$$= \pi c\omega X^2$$

In a similar line work done by spring and inertia forces can be written as follows:

$$W_s = \oint F_s\, dx \qquad\qquad F_s = kx$$

$$= \oint k\,x\,\dot{x}\,dt$$

$$= \oint k X^2\,\omega\,\sin(\omega t - \phi)\cos(\omega t - \phi)\,dt$$

$$= 0$$

$$W_I = \oint F_I\, dx \qquad\qquad F_I = m\ddot{x}$$

$$= \oint m\ddot{x}\,\dot{x}\,dt = \oint m\{-\omega^2 X\sin(\omega t - \phi)\}\{\omega X\cos(\omega t - \phi)\}$$

$$= -\omega^3 m X^2 \oint \sin(\omega t - \phi)\cos(\omega t - \phi)\,dt = 0$$

The energy dissipated per cycle by the damping force can be represented graphically as follows. Writing the velocity in the form

$$\dot{x} = \omega X \cos(\omega t - \varphi)$$

$$= \pm \omega X \sqrt{1 - \sin^2(\omega t - \varphi)}$$

$$= \pm \omega \sqrt{X^2 - x^2(t)}$$

The damping force becomes

$$F_d = c\dot{x} = \pm c\omega \sqrt{X^2 - x^2}$$

Rearranging the above equation to

$$\left(\frac{F_d}{c\omega X}\right)^2 + \left(\frac{x}{X}\right)^2 = 1$$

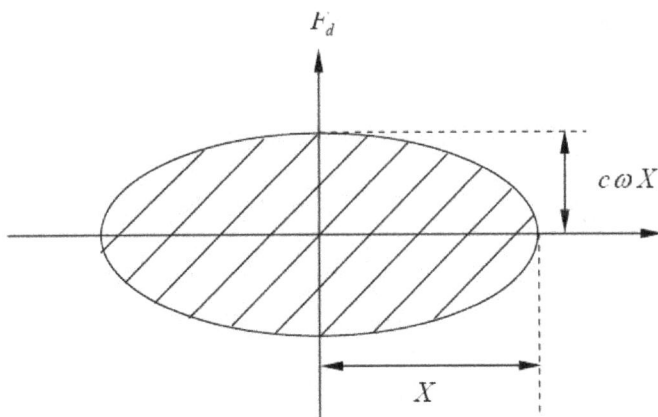

$$Area = \pi(c\omega X)(X) = \pi c\omega X^2$$

We recognize it as that of an ellipse with F_d and x plotted along the vertical and horizontal axis, as shown above. It is of interest to examine the total (elastic & damping) resisting force that is measured in an experiment.

$$F_d + F_s = c\dot{x} + kx$$
$$= kx + c\omega \sqrt{X^2 - x^2}$$

A plot of $F_s + F_d$ against x is the ellipse. The energy dissipated by damping is still the area enclosed by the ellipse.

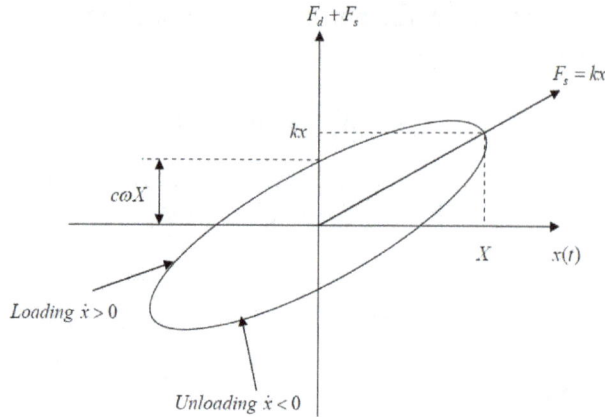

$$F_s = k\,x, is\ zero.$$

Damping properties of materials are listed in many different ways depending on the technical areas to which they are applied. Of these we list two relative energy units that have wide usage. First of these is specific damping capacity, defined as the energy loss per cycle W_d divided by the pick potential energy U.

Specific damping capacity $= \dfrac{E_d}{E_{so}}$

Where $E_{so} = \frac{1}{2}k\,X^2$

$\therefore$ Specific damping capacity $= \dfrac{\pi c\omega X^2}{\frac{1}{2}kX^2} = \dfrac{2\pi c\omega}{k}$

The second quantity is the loss coefficient or specific damping factor, is define as the ratio of damping energy loss per radian $w_d\big/2\pi$ divided by the peak potential or strain energy U.

$$\eta = \frac{w_d}{2\pi U} = \frac{c\omega}{k}$$

For the case of linear damping where the energy loss is proportional to the square of strain or amplitude, the hysteresis curve is an ellipse. When the damping loss is not quadratic function of the strain or amplitude, the hysteresis curve is no longer an ellipse.

Structural Damping

When materials are cyclically stressed, energy is dissipated internally within the material itself. Experiments by several investigators indicate that for most structural metals, such as steel, aluminium, the energy dissipated per cycle is independent of the frequency over a wide frequency range and proportional to the square of the amplitude of vibration. Internal damping fitting this classification is called solid damping or structural damping.

Energy dissipated by structural damping may be written as

$$w_d = \alpha X^2$$

Where α is constant with units of force / displacement. Using the concept of equivalent viscous damping gives

$$\pi C_{eq} \omega X^2 = \alpha X^2$$

or, $C_{eq} = \dfrac{\alpha}{\pi \omega}$

Substituting C_{eq} for c the differential equation of motion for a system with structural damping may be written as

$$m\ddot{x} + \left(\frac{\alpha}{\pi \omega}\right)\dot{x} + kx = F(t)$$

Complex stiffness: In the calculation of the flutter speed of air plane wings and tail surfaces, the concept of complex stiffness is used. It is arrived at by assuming the oscillations to be harmonic, which enables to be written as

$$m\ddot{x} + \left(k + i\frac{\alpha}{\pi}\right)x = F_0 e^{i\omega t}$$

Defining, $\gamma = \dfrac{\alpha}{k\pi}$ the equation becomes

$$m\ddot{x} + k(1+i\gamma)x = F_0 e^{i\omega t}$$

The quantity $k(1+i\gamma)$ is called complex stiffness and γ is the structural damping factor.

With the solution $x = X e^{i\omega t}$, the steady state amplitude becomes

$$X = \frac{F_0}{(k - m\omega^2) + i\gamma k}$$

The amplitude of resonance is then

$$|X| = \frac{F_0}{\gamma k}$$

Comparing this with the resonance amplitude of a system with viscous damping,

$$|X| = \frac{F_0}{2\zeta k} \qquad\qquad \gamma = 2\zeta$$

We conclude that with equal amplitudes at resonance, the structural damping factor is equal to twice the viscous damping factor.

Transient Loading

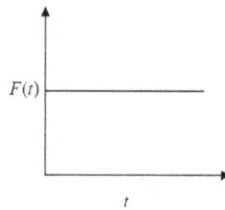

Suddenly applied step load   Gradually applied step load

When a dynamic system is excited by a suddenly applied non periodic excitation F(t) the response to such excitation is called transient response.

So far we have studied vibration of SDF system subject to steady state sinusoidal forces. However, the forces can have any general type of vibration with respect to time. Analyses of systems with such forces are often complicated. Presence of damping makes the problem more complicated. But all such problems can be solved without much difficulty through numerical method using digital computer.

I.   Suddenly applied load

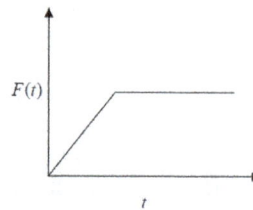

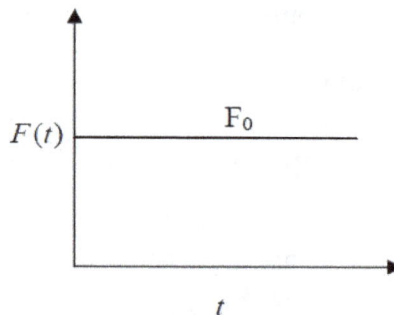

Eq. of motion

$$m\ddot{x} + k\,x = F_0$$

$$\Rightarrow \ddot{x} + \frac{k}{m}x = \frac{F_0}{m}$$

$$\Rightarrow \ddot{x} + \omega_n^2 \, x = \frac{F_0}{m}$$

Sol. Complementary function

$$x_c = A\sin \omega t + B\cos \omega$$

P. I $(D^2 + \omega^2)x \; = \; \dfrac{F_0}{m}$

$$x \;=\; \frac{F_0}{m}\cdot\frac{1}{(D^2 + \omega^2)} \;=\; \frac{F_0}{\omega^2 m}\left(1 + \frac{D}{\omega^2}\right)^{-1}$$

$$=\; \frac{F_0}{\omega^2 m}\left(1 - \frac{D^2}{\omega^2} + \ldots\ldots\right)$$

$$=\; \frac{F_0}{\omega^2 m} \;=\; \frac{F_0}{k}$$

$\therefore$ Complete solution ; $x \;=\; A\sin \omega t \;+\; B\cos \omega t \;+\; \dfrac{F_0}{k}$

P II $\dot{x} \;=\; A\omega\cos \omega t \;-\; B\omega\sin \omega t$

B. C. at $t = 0$ $x = 0$ $\dot{x} = 0$

$$0 = A.0 + B.1 + \frac{F_0}{k} \;\Rightarrow\; B = -\frac{F_0}{k}$$

$$0 = A\omega.1 - B.0 \quad \Rightarrow\; A = 0$$

$$\therefore \quad x \;=\; -\frac{F_0}{k}\cos \omega t + \frac{F_0}{k} \;=\; \frac{F_0}{k}(1-\cos \omega t) \;=\; x_{st}\,(1 - \cos \omega t)$$

$$\frac{x}{x_{st}} = dynamic \text{ amplification} = (1 - \cos \omega t)$$

$$\mathrm{x}_{max} = \frac{2F_0}{k} \quad \text{at} \quad \omega t = \pi$$

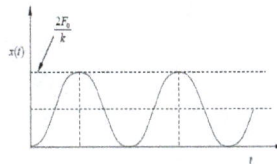

Response of SDOF System

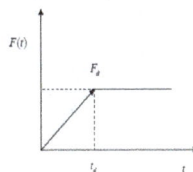

Eq. of motion

$$m\ddot{x} + kx = F_0 \frac{t}{t_d}$$

$$\Rightarrow \ddot{x} + \omega_n^2 \, x = \frac{F_0}{m} \cdot \frac{t}{t_d}$$

C.F.

$$x = A \sin \omega t + B \cos \omega t$$

P.I.

$$x = \frac{F_0}{m} \cdot \frac{1}{D^2 + \omega_n^2} \cdot \frac{t}{t_d}$$

$$= \frac{F_0}{m \omega_n^2} \cdot \left(1 + \frac{D^2}{\omega_n^2} \right) \cdot \frac{t}{t_d}$$

$$= \frac{F_0}{k} \cdot \frac{t}{t_d}$$

$$t \le t_d \quad F(t) = F_0 \frac{t}{t_d}$$

$$t \ge t_d \quad F(t) = F_0$$

Complete Solution:

$$x = A \sin \omega t + B \cos \omega t + \frac{F_0}{k} \cdot \frac{t}{t_d}$$

$$\text{B.C.} \quad t = 0 \quad x = 0 \quad \dot{x}(0) = 0 \quad B = 0$$

$$\dot{x} = A \omega \cos \omega t - B \sin \omega t + \frac{F_0}{k} \cdot \frac{1}{t_d}$$

$$x(0) = 0 = A \omega \cdot 1 - B \omega \cdot 0 + \frac{F_0}{k} \cdot \frac{1}{t_d}$$

$$\Rightarrow A = - \frac{F_0}{k \omega} \cdot \frac{1}{t_d}$$

$$\therefore \quad x(t) = - \frac{F_0}{k \omega t_d} \cdot \frac{1}{t_d} \sin \omega t + \frac{F_0}{k} \cdot \frac{t}{t_d}$$

$$= \frac{F_0}{k \omega t_d} \left[\omega t - \sin \omega t \right] = \frac{x_{st}}{\omega t_d} \left[\omega t - \sin \omega t \right]$$

$$\frac{x}{x_{st}} = Dynamic\ Magnification\ Factor \frac{1}{\omega t_d}[\omega t - \sin \omega t]$$

$$\dot{x} = \frac{F_0}{k\,\omega t_d}[\omega - \omega \cos \omega t] = \frac{F_0}{k\,t_d}[1 - \cos \omega t]$$

at $t = t_d$

$$x(t) = \frac{F_0}{k\,\omega t_d}\left[\omega t_d - \sin \omega t_d\right]$$

$$\dot{x}(t) = \frac{F_0}{k\,t_d}\left[1 - \cos \omega t_d\right]$$

When $t > t_d$ $m\ddot{x} + kx = F_0 \rightarrow earlier\ case$

$$x = A\sin \omega t + B\cos \omega t + \frac{F_0}{k}$$

$$\dot{x} = A\omega \cos \omega t - B\omega \sin \omega t$$

At $t = t_d$ initial conditions

Substituting initial conditions and solve for A & B to get the complete solution.

Initial conditions are

$$x(t_d) = A\sin \omega t_d + B\cos \omega t_d + \frac{F_0}{k} = \frac{F_0}{k\omega t_d}(\omega t_d - \sin \omega t_d)........ \quad (1)$$

$$\dot{x}(t_d) = A\omega \cos \omega t_d + B\omega \sin \omega t_d = \frac{F_0}{k\,t_d}(1 - \cos \omega t_d)........ \quad (2)$$

$$(2) \rightarrow A\cos \omega t_d - B\sin \omega t_d = \frac{F_0}{k\omega t_d}(1 - \cos \omega t_d)$$

$$(1) \times \sin \omega t_d \Rightarrow A\sin^2 \omega t_d + B\sin \omega t_d \cos \omega t_d = \frac{F_0 \sin \omega t_d}{k\omega t_d}(\omega t_d - \sin \omega t_d)$$

$$-\frac{F_0 \sin \omega t_d}{k}$$

$$(2)\times\cos\omega\,t_d \Rightarrow A\cos^2\omega t_d - B\sin\omega t_d\cos\omega t_d = \frac{F_0\cos\omega t_d}{kt_d}\left(1-\cos\omega t_d\right)$$

Adding above two equations we get

$$S\left(\sin^2\omega t_d+\cos^2\omega t_d\right)=\frac{F_0}{k\omega t_d}\left(\omega t_d\sin\omega t_d-\sin^2\omega t_d+\cos\omega t_d-\sin^2\omega t_d\right)-\frac{F_0}{k}\sin\omega t_d$$

$$\Rightarrow A=\frac{F_0}{k\omega t_d}\left(\cos\omega t_d+\omega t_d\sin\omega t_d-1\right)-\frac{F_0}{k}\sin\omega t_d$$

Similarly, multiplying equation (1) by d $\cos\omega t_d$ & (2) by d $\sin\omega t_d$ we get,

$$1)\times\cos\omega t_d \Rightarrow A\sin\omega t_d\cos\omega t_d+B\cos^2\omega t_d=\frac{F_0\cos\omega t_d}{k\omega t_d}\left(\omega t_d-\sin\omega t_d\right)$$

$$-\frac{F_0\cos\omega t_d}{k}$$

$$(2)\times\sin\omega t_d \Rightarrow A\cos\omega t_d\sin\omega t_d-B\sin^2\omega t_d=\frac{F_0\sin\omega t_d}{kt_d}\left(1-\cos\omega t_d\right)$$

substracting (2) from (1) we get,

$$B\left(\sin^2\omega t_d+\cos^2\omega t_d\right)$$

$$\frac{F_0}{k\omega t_d}\left(\omega t_d\cos\omega t_d-\sin\omega t_d\cos\omega t_d-\sin\omega t_d+\sin\omega t_d\cos\omega t_d\right)-\frac{F_0}{k}\cos\omega t$$

$$\Rightarrow B=\frac{F_0}{k\omega t_d}\left(\omega t_d\cos\omega t_d-\sin\omega t_d\right)-\frac{F_0}{k}\cos\omega t$$

∴ Response at any time $t>t_d$ we get,

$$x(t)=\left\{\frac{F_0}{k\omega t_d}\left(\cos\omega t_d+\omega t_d\sin\omega t_d-1\right)-\frac{F_0}{k}\sin\omega t_d\right\}\sin\omega t$$

$$+\left\{\frac{F_0}{k\omega t_d}\left(\omega t_d\cos\omega t_d-\sin\omega t_d\right)-\frac{F_0}{k}\cos\omega t\right\}\cos\omega t+\frac{F_0}{k}$$

The dynamic loading resulting from blast, gust of wind or seismic forces is generally not harmonic. In such cases the equation of motions has to be solved numerically. Solution in exact form can be obtain only for some idealized loading conditions.

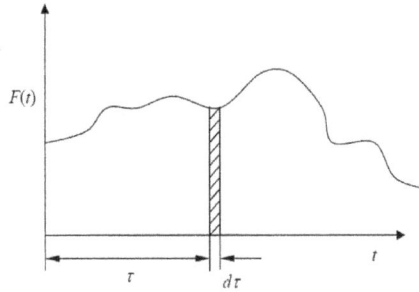

Consider an undamped, SDF system subject to a disturbing force F(t) which is a function of time. The solution of problem can be obtained by the superposition of individual results. We shall first consider the effect of a single response subjected to a force $F(\tau)$. From impulse momentum equation i.e. $m\dot{x} = F(\tau)d\tau$, the impulse acting on the mass will result in a sudden change in its velocity equal to $\dot{x} = \dfrac{F(\tau)d\tau}{m}$ without an appreciable change in displacement.

Under free vibration we found that undamped spring-mass system with initial conditions $x(0)\, and\, \dot{x}(0)$:

$$x(t) = \frac{\dot{x}(0)}{\omega_n}\sin \omega_n t + x(0)\cos \omega_n t$$

The displacement at any time t with initial velocity can be written as

$$x(t) = \frac{\dot{x}(0)}{\omega_n}\sin \omega_n t \frac{F(\tau)\sin \omega_n t}{m\omega_n}d\tau$$

Having obtained the displacement under the effect of an impulse load, we may consider the displacement after application of the impulse given by the above equation except the increase of t we must put $(t-\tau)$, then

$$x(t) = \frac{F(\tau)\sin \omega_n (t-\tau)d\tau}{m\omega_n}$$

∴ Total displacement due to the arbitrary load applied to the SDF can be written as

$$x(t) = \int_0^t \frac{F(\tau)}{m\omega_n}\sin \omega_n (t-\tau)d\tau$$

Duhamal's integral

Convolution integral

Green's Function

When damping is present we can start with the free vibration equation, with $x(0) = 0$ and displacement with single impulse:

$$x(t) = \frac{\dot{x}(0) e^{-\zeta \omega_n t}}{\omega_n \sqrt{1-\zeta^2}} \sin \sqrt{1-\zeta^2}\, \omega_n t$$

Total displacement if damping is considered

$$x(t) = \frac{1}{m \omega_n \sqrt{1-\zeta^2}} \int_0^t e^{-\zeta \omega_n (t-\tau)} F(\tau) \sin\left[\omega_n \sqrt{1-\zeta^2}\,(t-\tau)\right] d\tau$$

The above expression can be used directly for computing the response of SDF system of any forcing function provided the integral is to be evaluated.

Prob. 1

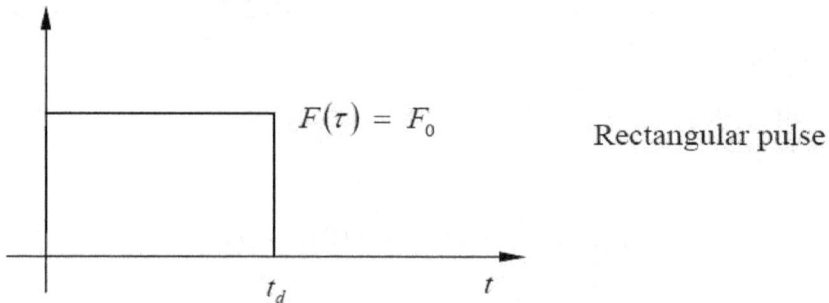

$F(\tau) = F_0$ Rectangular pulse

$t > t_d$ *free vibration part*

$t < t_d$ *forced vibration part*

$$x(t) = \frac{1}{m \omega_n} \int_0^t F(\tau) \sin \omega_n (t-\tau) d\tau$$

$$= \frac{1}{m \omega_n} \int_0^t F_0 \sin \omega_n (t-\tau) d\tau$$

$$= \frac{1}{m \omega_n} \left[+ \frac{F_0}{\omega_n} \cos \omega_n (t-\tau) \right]_0^t$$

$$= \frac{1}{m \omega_n^2} \left[1 - \cos \omega_n t \right]$$

$$= \frac{F_0}{k}\left(1 - \cos \omega_n t\right)$$

For $t > t_d$, the free vibration response can be obtained after substituting velocity displacement at $t = t_d$

$$x(t) = x_{td} \cos(t - t_d)\omega_n + \frac{\dot{x}_{td}}{\omega_n} \sin \omega_n (t - t_d)$$

$$t \leq t_d \qquad\qquad t = t_d$$

$$x(t) = \frac{F_0}{k}\left(1 - \cos \omega_n t\right) \qquad x(t_d) = \frac{F_0}{k}\left(1 - \cos \omega_n t_d\right)$$

$$\dot{x}(t) = \frac{F_0}{k} \omega_n \sin \omega_n t \qquad \dot{x}(t_d) = \frac{F_0}{k} \omega_n \sin \omega_n t_d$$

$t > t_d$ the response can be written as

$$x(t) = x_0 \cos \omega_n t + \frac{\dot{x}_0}{\omega_n} \sin \omega_n t$$

$$= \frac{F_0}{k}\left(1 - \cos \omega_n t_d\right) \cos \omega_n (t - t_d) + \frac{F_0}{k} \sin \omega_n t_d \sin \omega_n (t - t_d)$$

$$= \frac{F_0}{k}\left[\cos \omega_n (t - t_d) - \cos \omega_n t\right] \rightarrow simplifing$$

$$\text{Dynamic Load Factor (DLF)} = \frac{\text{Displacement at any time } t}{\text{Static displacement}\left(x_{st} = \dfrac{F_0}{k}\right)}$$

$$= 1 - \cos \omega_n t \qquad t \leq t_d$$

$$= \cos \omega_n (t - t_d) - \cos \omega_n t_d \qquad t > t_d$$

For $t > 1$

$$x(t) = \int_0^1 \frac{F_0}{m \omega_n} \sin \omega_n (t - t_d) d\tau = \frac{F_0}{m \omega_n} \cos \omega_n (t - t_d)\Big|_0^1$$

$$= \frac{F_0}{k} \left[\cos \omega_n (t-1) - \cos \omega_n t \right]$$

Prob. 2

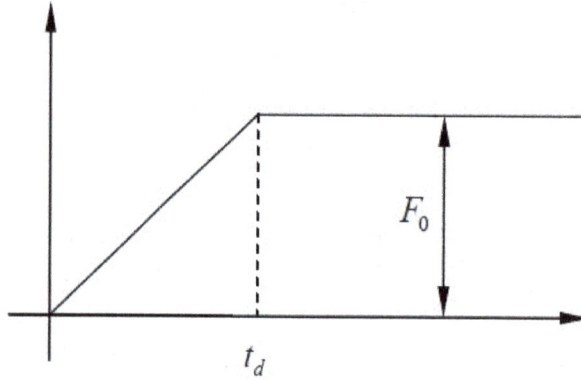

For first ramp function

$$x(t) = \frac{F_0}{m \omega_n} \int_0^t F_0 \left(\frac{t}{t_d} \right) \sin \omega_n (t - \tau) d\tau$$

$$= \frac{F_0}{k} \left(\frac{t}{t_d} - \frac{\cos \omega_n t}{\omega_n t_d} \right) \qquad t < t_d$$

For the second ramp function starting at t_d the solution can be written by inspection of the above equation as

$$x(t) = -\frac{F_0}{k} \left[\frac{t - t_d}{t_d} - \frac{\cos \omega_n (t - t_d)}{\omega_n t_d} \right]$$

By superposition of these two equations the response for $t > t_d$ becomes

$$x(t) = \frac{F_0}{k} \left[1 - \frac{\cos \omega_{nt}}{\omega_n t_d} + \frac{\cos \omega_n (t - t_d)}{\omega_n t_d} \right]$$

Laplace Transform

In mathematics the Laplace transform is an integral transform named after its discoverer Pierre-Simon Laplace. It takes a function of a real variable t (often time) to a function of a complex variable s (frequency).

The Laplace transform is very similar to the Fourier transform. While the Fourier transform of a function is a complex function of a *real* variable (frequency), the Laplace transform of a function is a complex function of a *complex variable*. Laplace transforms are usually restricted to functions of t with $t > 0$. A consequence of this restriction is that the Laplace transform of a function is a holomorphic function of the variable s. Unlike the Fourier transform, the Laplace transform of a distribution is generally a well-behaved function. Also techniques of complex variables can be used directly to study Laplace transforms. As a holomorphic function, the Laplace transform has a power series representation. This power series expresses a function as a linear superposition of moments of the function. This perspective has applications in probability theory.

The Laplace transform is invertible on a large class of functions. The inverse Laplace transform takes a function of a complex variable s (often frequency) and yields a function of a real variable t (time). Given a simple mathematical or functional description of an input or output to a system, the Laplace transform provides an alternative functional description that often simplifies the process of analyzing the behavior of the system, or in synthesizing a new system based on a set of specifications. So, for example, Laplace transformation from the time domain to the frequency domain transforms differential equations into algebraic equations and convolution into multiplication. It has many applications in the sciences and technology.

History

The Laplace transform is named after mathematician and astronomer Pierre-Simon Laplace, who used a similar transform (now called the z-transform) in his work on probability theory. The current widespread use of the transform (mainly in engineering) came about during and soon after World War II although it had been used in the 19th century by Abel, Lerch, Heaviside, and Bromwich.

The early history of methods having some similarity to Laplace transform is as follows. From 1744, Leonhard Euler investigated integrals of the form

$$z = \int X(x)e^{ax}\,dx \quad \text{and} \quad z = \int X(x)x^A\,dx$$

as solutions of differential equations but did not pursue the matter very far.

Joseph Louis Lagrange was an admirer of Euler and, in his work on integrating probability density functions, investigated expressions of the form

$$\int X(x)e^{-ax}a^x\,dx,$$

which some modern historians have interpreted within modern Laplace transform theory.

These types of integrals seem first to have attracted Laplace's attention in 1782 where he was following in the spirit of Euler in using the integrals themselves as solutions of equations. However, in 1785, Laplace took the critical step forward when, rather than just looking for a solution in the form of an integral, he started to apply the transforms in the sense that was later to become popular. He used an integral of the form

$$\int x^s \varphi(x)dx,$$

akin to a Mellin transform, to transform the whole of a difference equation, in order to look for solutions of the transformed equation. He then went on to apply the Laplace transform in the same way and started to derive some of its properties, beginning to appreciate its potential power.

Laplace also recognised that Joseph Fourier's method of Fourier series for solving the diffusion equation could only apply to a limited region of space because those solutions were periodic. In 1809, Laplace applied his transform to find solutions that diffused indefinitely in space.

Formal Definition

The Laplace transform is a frequency-domain approach for continuous time signals irrespective of whether the system is stable or unstable. The Laplace transform of a function $f(t)$, defined for all real numbers $t \geq 0$, is the function $F(s)$, which is a unilateral transform defined by

$$F(s) = \int_0^\infty f(t)e^{-st}\,dt$$

where s is a complex number frequency parameter

$s = \sigma + i\omega,$, with real numbers σ and ω.

Other notations for the Laplace transform include $L\{f\}$, or alternatively $L\{f(t)\}$ instead of F.

The meaning of the integral depends on types of functions of interest. A necessary condition for existence of the integral is that f must be locally integrable on $[0, \infty)$. For locally integrable functions that decay at infinity or are of exponential type, the integral can be understood to be a (proper) Lebesgue integral. However, for many applications it is necessary to regard it to be a conditionally convergent improper integral at ∞. Still more generally, the integral can be understood in a weak sense, and this is dealt with below.

One can define the Laplace transform of a finite Borel measure μ by the Lebesgue integral

$$\mathcal{L}\{\mu\}(s) = \int_{[0,\infty)} e^{-st}\, d\mu(t).$$

An important special case is where μ is a probability measure, for example, the Dirac delta function. In operational calculus, the Laplace transform of a measure is often treated as though the measure came from a probability density function f. In that case, to avoid potential confusion, one often writes

$$\mathcal{L}\{f\}(s) = \int_{0^-}^{\infty} f(t)e^{-st}\, dt,$$

where the lower limit of 0^- is shorthand notation for

$$\lim_{\varepsilon \downarrow 0} \int_{-\varepsilon}^{\infty}.$$

This limit emphasizes that any point mass located at 0 is entirely captured by the Laplace transform. Although with the Lebesgue integral, it is not necessary to take such a limit, it does appear more naturally in connection with the Laplace–Stieltjes transform.

Probability Theory

In pure and applied probability, the Laplace transform is defined as an expected value. If X is a random variable with probability density function f, then the Laplace transform of f is given by the expectation

$$\mathcal{L}\{f\}(s) = E\left[e^{-sX}\right].$$

By abuse of language, this is referred to as the Laplace transform of the random variable X itself. Replacing s by $-t$ gives the moment generating function of X. The Laplace transform has applications throughout probability theory, including first passage times of stochastic processes such as Markov chains, and renewal theory.

Of particular use is the ability to recover the cumulative distribution function of a continuous random variable X by means of the Laplace transform as follows

$$F_X(x) = \mathcal{L}^{-1}\left\{\frac{1}{s}E\left[e^{-sX}\right]\right\}(x) = \mathcal{L}^{-1}\left\{\frac{1}{s}\mathcal{L}\{f\}(s)\right\}(x).$$

Bilateral Laplace Transform

When one says "the Laplace transform" without qualification, the unilateral or one-sided transform is normally intended. The Laplace transform can be alternatively defined as the *bilateral Laplace transform* or two-sided Laplace transform by extending the limits of integration to be the entire real axis. If that is done the common unilateral transform simply becomes a special case of the bilateral transform where the definition of the function being transformed is multiplied by the Heaviside step function.

The bilateral Laplace transform is defined as follows,

$$\mathcal{B}\{f\}(s) = \int_{-\infty}^{\infty} e^{-st} f(t)dt.$$

Inverse Laplace Transform

Two integrable functions have the same Laplace transform only if they differ on a set of Lebesgue measure zero. This means that, on the range of the transform, there is an inverse transform. In fact, besides integrable functions, the Laplace transform is a one-to-one mapping from one function space into another in many other function spaces as well, although there is usually no easy characterization of the range. Typical function spaces in which this is true include the spaces of bounded continuous functions, the space $L^{\infty}(0, \infty)$, or more generally tempered functions (that is, functions of at worst polynomial growth) on $(0, \infty)$. The Laplace transform is also defined and injective for suitable spaces of tempered distributions.

In these cases, the image of the Laplace transform lives in a space of analytic functions in the region of convergence. The inverse Laplace transform is given by the following complex integral, which is known by various names (the Bromwich integral, the Fourier–Mellin integral, and Mellin's inverse formula):

$$f(t) = \mathcal{L}^{-1}\{F\}(t) = \frac{1}{2\pi i} \lim_{T\to\infty} \int_{\gamma-iT}^{\gamma+iT} e^{st} F(s)ds,$$

where γ is a real number so that the contour path of integration is in the region of convergence of $F(s)$. An alternative formula for the inverse Laplace transform is given by Post's inversion formula. The limit here is interpreted in the weak-* topology.

In practice, it is typically more convenient to decompose a Laplace transform into known transforms of functions obtained from a table, and construct the inverse by inspection.

Region of Convergence

If f is a locally integrable function (or more generally a Borel measure locally of bounded variation), then the Laplace transform $F(s)$ of f converges provided that the limit exists.

$$\lim_{R \to \infty} \int_0^R f(t)e^{-st}\, dt$$

The Laplace transform converges absolutely if the integral

$$\int_0^\infty \left| f(t)e^{-st} \right| dt$$

exists (as a proper Lebesgue integral). The Laplace transform is usually understood as conditionally convergent, meaning that it converges in the former instead of the latter sense.

The set of values for which $F(s)$ converges absolutely is either of the form $\mathrm{Re}(s) > a$ or else $\mathrm{Re}(s) \geq a$, where a is an extended real constant, $-\infty \leq a \leq \infty$. (This follows from the dominated convergence theorem.) The constant a is known as the abscissa of absolute convergence, and depends on the growth behavior of $f(t)$. Analogously, the two-sided transform converges absolutely in a strip of the form $a < \mathrm{Re}(s) < b$, and possibly including the lines $\mathrm{Re}(s) = a$ or $\mathrm{Re}(s) = b$. The subset of values of s for which the Laplace transform converges absolutely is called the region of absolute convergence or the domain of absolute convergence. In the two-sided case, it is sometimes called the strip of absolute convergence. The Laplace transform is analytic in the region of absolute convergence.

Similarly, the set of values for which $F(s)$ converges (conditionally or absolutely) is known as the region of conditional convergence, or simply the region of convergence (ROC). If the Laplace transform converges (conditionally) at $s = s_0$, then it automatically converges for all s with $\mathrm{Re}(s) > \mathrm{Re}(s_0)$. Therefore, the region of convergence is a half-plane of the form $\mathrm{Re}(s) > a$, possibly including some points of the boundary line $\mathrm{Re}(s) = a$.

In the region of convergence $\mathrm{Re}(s) > \mathrm{Re}(s_0)$, the Laplace transform of f can be expressed by integrating by parts as the integral

$$F(s) = (s - s_0)\int_0^\infty e^{-(s-s_0)t}\beta(t)\, dt, \quad \beta(u) = \int_0^u e^{-s_0 t} f(t)\, dt.$$

That is, in the region of convergence $F(s)$ can effectively be expressed as the absolutely convergent Laplace transform of some other function. In particular, it is analytic.

There are several Paley–Wiener theorems concerning the relationship between the decay properties of f and the properties of the Laplace transform within the region of convergence.

In engineering applications, a function corresponding to a linear time-invariant (LTI) system is *stable* if every bounded input produces a bounded output. This is equivalent to the absolute convergence of the Laplace transform of the impulse response function in the region $\text{Re}(s) \geq 0$. As a result, LTI systems are stable provided the poles of the Laplace transform of the impulse response function have negative real part.

This ROC is used in knowing about the causality and stability of a system.

Properties and Theorems

The Laplace transform has a number of properties that make it useful for analyzing linear dynamical systems. The most significant advantage is that differentiation and integration become multiplication and division, respectively, by s (similarly to logarithms changing multiplication of numbers to addition of their logarithms).

Because of this property, the Laplace variable s is also known as *operator variable* in the L domain: either *derivative operator* or (for s^{-1}) *integration operator*. The transform turns integral equations and differential equations to polynomial equations, which are much easier to solve. Once solved, use of the inverse Laplace transform reverts to the time domain.

Given the functions $f(t)$ and $g(t)$, and their respective Laplace transforms $F(s)$ and $G(s)$,

$$g(t) = \mathcal{L}^{-1}\{G(s)\},$$
$$f(t) = \mathcal{L}^{-1}\{F(s)\},$$

The following table is a list of properties of unilateral Laplace transform:

Properties of the unilateral Laplace transform			
	Time domain	**s domain**	**Comment**
Linearity	$af(t) + bg(t)$	$aF(s) + bG(s)$	Can be proved using basic rules of integration.
Frequency-domain derivative	$tf(t)$	$-F'(s)$	F' is the first derivative of F with respect to s.
Frequency-domain general derivative	$t^n f(t)$	$(-1)^n F^{(n)}(s)$	More general form, nth derivative of $F(s)$.
Derivative	$f'(t)$	$sF(s) - f(0)$	f is assumed to be a differentiable function, and its derivative is assumed to be of exponential type. This can then be obtained by integration by parts

Second derivative	$f''(t)$	$s^2 F(s) - sf(0) - f'(0)$	f is assumed twice differentiable and the second derivative to be of exponential type. Follows by applying the Differentiation property to $f'(t)$.
General derivative	$f^{(n)}(t)$	$s^n F(s) - \sum_{k=1}^{n} s^{n-k} f^{(k-1)}(0)$	f is assumed to be n-times differentiable, with nth derivative of exponential type. Follows by mathematical induction.
Frequency-domain integration	$\dfrac{1}{t} f(t)$	$\displaystyle\int_s^\infty F(\sigma)d\sigma$	This is deduced using the nature of frequency differentiation and conditional convergence.
Time-domain integration	$\displaystyle\int_0^t f(\tau)d\tau = (u * f)(t)$	$\dfrac{1}{s} F(s)$	$u(t)$ is the Heaviside step function and $(u * f)(t)$ is the convolution of $u(t)$ and $f(t)$.
Frequency shifting	$e^{at} f(t)$	$F(s-a)$	
Time shifting	$f(t-a)u(t-a)$	$e^{-as} F(s)$	$u(t)$ is the Heaviside step function
Time scaling	$f(at)$	$\dfrac{1}{a} F\left(\dfrac{s}{a}\right)$	
Multiplication	$f(t)g(t)$	$\dfrac{1}{2\pi i} \lim_{T\to\infty} \int_{c-iT}^{c+iT} F(\sigma)G(s-\sigma)d\sigma$	The integration is done along the vertical line $\mathrm{Re}(\sigma) = c$ that lies entirely within the region of convergence of F.
Convolution	$(f * g)(t) = \displaystyle\int_0^t f(\tau)g(t-\tau)d\tau$	$F(s) \cdot G(s)$	
Complex conjugation	$f^*(t)$	$F^*(s^*)$	

Cross-cor-relation	$f(t) \star g(t)$	$F^*(-s^*) \cdot G(s)$	
Periodic function	$f(t)$	$\dfrac{1}{1-e^{-Ts}} \displaystyle\int_0^T e^{-st} f(t)dt$	$f(t)$ is a periodic function of period T so that $f(t) = f(t + T)$, for all $t \geq 0$. This is the result of the time shifting property and the geometric series.

- Initial value theorem:

$$f(0^+) = \lim_{s \to \infty} sF(s).$$

- Final value theorem:

$f(\infty) = \lim_{s \to 0} sF(s)$, if all poles of $sF(s)$ are in the left half-plane.

The final value theorem is useful because it gives the long-term behaviour without having to perform partial fraction decompositions or other difficult algebra. If $F(s)$ has a pole in the right-hand plane or poles on the imaginary axis (e.g., if $f(t) = e^t$ or $f(t) = \sin(t)$), the behaviour of this formula is undefined.

Relation to Power Series

The Laplace transform can be viewed as a continuous analogue of a power series. If $a(n)$ is a discrete function of a positive integer n, then the power series associated to $a(n)$ is the series

$$\sum_{n=0}^{\infty} a(n)x^n$$

where x is a real variable. Replacing summation over n with integration over t, a continuous version of the power series becomes

$$\int_0^{\infty} f(t)x^t \, dt$$

where the discrete function $a(n)$ is replaced by the continuous one $f(t)$.

Changing the base of the power from x to e gives

$$\int_0^{\infty} f(t)\left(e^{\ln x}\right)^t dt$$

For this to converge for, say, all bounded functions f, it is necessary to require that $\ln x < 0$. Making the substitution $-s = \ln x$ gives just the Laplace transform:

$$\int_0^\infty f(t)e^{-st}\,dt$$

In other words, the Laplace transform is a continuous analog of a power series in which the discrete parameter n is replaced by the continuous parameter t, and x is replaced by e^{-s}.

Relation to Moments

The quantities

$$\mu_n = \int_0^\infty t^n f(t)\,dt$$

are the *moments* of the function f. If the first n moments of f converge absolutely, then by repeated differentiation under the integral,

$$(-1)^n (\mathcal{L}f)^{(n)}(0) = \mu_n.$$

This is of special significance in probability theory, where the moments of a random variable X are given by the expectation values $\mu_n = E[X^n]$. Then, the relation holds

$$\mu_n = (-1)^n \frac{d^n}{ds^n} E\left[e^{-sX}\right](0).$$

Proof of the Laplace Transform of a Function's Derivative

It is often convenient to use the differentiation property of the Laplace transform to find the transform of a function's derivative. This can be derived from the basic expression for a Laplace transform as follows:

$$\mathcal{L}\{f(t)\} = \int_{0^-}^\infty e^{-st} f(t)\,dt$$

$$= \left[\frac{f(t)e^{-st}}{-s}\right]_{0^-}^\infty - \int_{0^-}^\infty \frac{e^{-st}}{-s} f'(t)\,dt \quad \text{(by parts)}$$

$$= \left[-\frac{f(0^-)}{-s}\right] + \frac{1}{s}\mathcal{L}\{f'(t)\},$$

yielding

$$\mathcal{L}\{f'(t)\} = s\cdot\mathcal{L}\{f(t)\} - f(0^-),$$

and in the bilateral case,

$$\mathcal{L}\{f'(t)\} = s \int_{-\infty}^{\infty} e^{-st} f(t)dt = s \cdot \mathcal{L}\{f(t)\}.$$

The general result

$$\mathcal{L}\{f^{(n)}(t)\} = s^n \cdot \mathcal{L}\{f(t)\} - s^{n-1} f(0^-) - \cdots - f^{(n-1)}(0^-),$$

where $f^{(n)}$ denotes the nth derivative of f, can then be established with an inductive argument.

Evaluating Integrals Over the Positive Real Axis

A really useful property of the Laplace transform is the following:

$$\int_0^{+\infty} f(x)g(x)dx = \int_0^{+\infty} (\mathcal{L}f)(s) \cdot (\mathcal{L}^{-1}g)(s)ds$$

under suitable assumptions on the behaviour of f, g in a right neighbourhood of 0 and on the decay rate of f, g in a left neighbourhood of ∞. The above formula is a variation of integration by parts, with the operators $\dfrac{d}{dx}$ and $\int dx$ being replaced by $\mathcal{L}$ and $\mathcal{L}^{-1}$. Let us prove the equivalent formulation:

$$\int_0^{+\infty} (\mathcal{L}f)(x)g(x)dx = \int_0^{+\infty} f(s)(\mathcal{L}g)(s)ds.$$

By plugging in $(\mathcal{L}f)(x) = \int_0^{+\infty} f(s)e^{-sx} ds$ the left-hand side turns into:

$$\int_0^{+\infty} \int_0^{+\infty} f(s)g(x)e^{-sx} ds\,dx,$$

but assuming Fubini's theorem holds, by reversing the order of integration we get the wanted right-hand side.

Evaluating Improper Integrals

Let $\mathcal{L}\{f(t)\} = F(s)$, then

$$\mathcal{L}\left\{ \frac{f(t)}{t} \right\} = \int_s^{\infty} F(p)dp,$$

or

$$\int_0^\infty \frac{f(t)}{t} e^{-st} dt = \int_s^\infty F(p)dp.$$

Letting $s \to 0$, gives one the identity

$$\int\limits_0^\infty \frac{f(t)}{t} dt = \int\limits_0^\infty F(p)dp.$$

provided that the interchange of limits can be justified. Even when the interchange cannot be justified the calculation can be suggestive. For example, proceeding formally one has

$$\int\limits_0^\infty \frac{1}{t}(\cos(at) - \cos(bt))dt = \int\limits_0^\infty \left(\frac{p}{p^2 + a^2} - \frac{p}{p^2 + b^2} \right) dp$$

$$= \frac{1}{2} \ln \frac{p^2 + a^2}{p^2 + b^2} \bigg|_{p:=0}^\infty = \ln b - \ln a.$$

The validity of this identity can be proved by other means. It is an example of a Frullani integral.

Another example is Dirichlet integral.

Relationship to Other Transforms

Laplace–Stieltjes Transform

The (unilateral) Laplace–Stieltjes transform of a function $g : R \to R$ is defined by the Lebesgue–Stieltjes integral

$$\{\mathcal{L}^* g\}(s) = \int\limits_0^\infty e^{-st} dg(t).$$

The function g is assumed to be of bounded variation. If g is the antiderivative of f:

$$g(x) = \int_0^x f(t)dt$$

then the Laplace–Stieltjes transform of g and the Laplace transform of f coincide. In general, the Laplace–Stieltjes transform is the Laplace transform of the Stieltjes measure associated to g. So in practice, the only distinction between the two transforms is that the Laplace transform is thought of as operating on the density function of the measure, whereas the Laplace–Stieltjes transform is thought of as operating on its cumulative distribution function.

Fourier Transform

The continuous Fourier transform is equivalent to evaluating the bilateral Laplace transform with imaginary argument $s = i\omega$ or $s = 2\pi fi$,

$$\hat{f}(\omega) = \mathcal{F}\{f(t)\}$$

$$= \mathcal{L}\{f(t)\}|_{s=i\omega} = F(s)|_{s=i\omega}$$

$$= \int_{-\infty}^{\infty} e^{-i\omega t} f(t)dt \, .$$

This definition of the Fourier transform requires a prefactor of $1/2\,\pi$ on the reverse Fourier transform. This relationship between the Laplace and Fourier transforms is often used to determine the frequency spectrum of a signal or dynamical system.

The above relation is valid as stated if and only if the region of convergence (ROC) of $F(s)$ contains the imaginary axis, $\sigma = 0$.

For example, the function $f(t) = \cos(\omega_0 t)$ has a Laplace transform $F(s) = s/(s^2 + \omega_0^2)$ whose ROC is $\mathrm{Re}(s) > 0$. As $s = i\omega$ is a pole of $F(s)$, substituting $s = i\omega$ in $F(s)$ does not yield the Fourier transform of $f(t)u(t)$, which is proportional to the Dirac delta-function $\delta(\omega - \omega_0)$.

However, a relation of the form

$$\lim_{\sigma \to 0^+} F(\sigma + i\omega) = \hat{f}(\omega)$$

holds under much weaker conditions. For instance, this holds for the above example provided that the limit is understood as a weak limit of measures. General conditions relating the limit of the Laplace transform of a function on the boundary to the Fourier transform take the form of Paley–Wiener theorems.

Mellin Transform

The Mellin transform and its inverse are related to the two-sided Laplace transform by a simple change of variables.

If in the Mellin transform

$$G(s) = \mathcal{M}\{g(\theta)\} = \int_0^\infty \theta^s g(\theta) \frac{d\theta}{\theta}$$

we set $\theta = e^{-t}$ we get a two-sided Laplace transform.

Z-transform

The unilateral or one-sided Z-transform is simply the Laplace transform of an ideally sampled signal with the substitution of

$$z \overset{\text{def}}{=} e^{sT},$$

where $T = 1/f_s$ is the sampling period (in units of time e.g., seconds) and f_s is the sampling rate (in samples per second or hertz).

Let

$$\Delta_T(t) \overset{\text{def}}{=} \sum_{n=0}^\infty \delta(t - nT)$$

be a sampling impulse train (also called a Dirac comb) and

$$x_q(t) \overset{\text{def}}{=} x(t)\Delta_T(t) = x(t)\sum_{n=0}^\infty \delta(t - nT)$$

$$= \sum_{n=0}^\infty x(nT)\delta(t - nT) = \sum_{n=0}^\infty x[n]\delta(t - nT)$$

be the sampled representation of the continuous-time $x(t)$

$$x[n] \overset{\text{def}}{=} x(nT) .$$

The Laplace transform of the sampled signal $x_{q(t)}$ is

$$X_q(s) = \int_{0^-}^\infty x_q(t)e^{-st}\, dt$$

$$= \int_{0^-}^\infty \sum_{n=0}^\infty x[n]\delta(t - nT)e^{-st}\, dt$$

$$dt = \sum_{n=0}^\infty x[n] \int_{0^-}^\infty \delta(t - nT)e^{-st}\, dt$$

$$= \sum_{n=0}^{\infty} x[n] e^{-nsT} .$$

This is the precise definition of the unilateral Z-transform of the discrete function $x[n]$

$$X(z) = \sum_{n=0}^{\infty} x[n] z^{-n}$$

with the substitution of $z \rightarrow e^{sT}$.

Comparing the last two equations, we find the relationship between the unilateral Z-transform and the Laplace transform of the sampled signal,

$$X_q(s) = X(z) \big|_{z=e^{sT}} .$$

The similarity between the Z and Laplace transforms is expanded upon in the theory of time scale calculus.

Borel Transform

The integral form of the Borel transform

$$F(s) = \int_0^{\infty} f(z) e^{-sz} dz$$

is a special case of the Laplace transform for f an entire function of exponential type, meaning that

$$|f(z)| \le A e^{B|z|}$$

for some constants A and B. The generalized Borel transform allows a different weighting function to be used, rather than the exponential function, to transform functions not of exponential type. Nachbin's theorem gives necessary and sufficient conditions for the Borel transform to be well defined.

Fundamental Relationships

Since an ordinary Laplace transform can be written as a special case of a two-sided transform, and since the two-sided transform can be written as the sum of two one-sided transforms, the theory of the Laplace-, Fourier-, Mellin-, and Z-transforms are at bottom the same subject. However, a different point of view and different characteristic problems are associated with each of these four major integral transforms.

Table of Selected Laplace Transforms

The following table provides Laplace transforms for many common functions of a single variable.

Because the Laplace transform is a linear operator,

- The Laplace transform of a sum is the sum of Laplace transforms of each term.

$$\mathcal{L}\{f(t)+g(t)\} = \mathcal{L}\{f(t)\} + \mathcal{L}\{g(t)\}$$

- The Laplace transform of a multiple of a function is that multiple times the Laplace transformation of that function.

$$\mathcal{L}\{af(t)\} = a\mathcal{L}\{f(t)\}$$

Using this linearity, and various trigonometric, hyperbolic, and complex number (etc.) properties and/or identities, some Laplace transforms can be obtained from others quicker than by using the definition directly.

The unilateral Laplace transform takes as input a function whose time domain is the non-negative reals, which is why all of the time domain functions in the table below are multiples of the Heaviside step function, $u(t)$.

The entries of the table that involve a time delay τ are required to be causal (meaning that $\tau > 0$). A causal system is a system where the impulse response $h(t)$ is zero for all time t prior to $t = 0$. In general, the region of convergence for causal systems is not the same as that of anticausal systems.

Function	Time domain $f(t) = \mathcal{L}^{-1}\{F(s)\}$	Laplace S-domain $F(s) = \mathcal{L}\{f(t)\}$	Region of convergence
unit impulse	$\delta(t)$	1	all s
delayed impulse	$\delta(t-\tau)$	$e^{-\tau s}$	
unit step	$u(t)$	$\dfrac{1}{s}$	$Re(s) > 0$
delayed unit step	$u(t-\tau)$	$\dfrac{1}{s}e^{-\tau s}$	$Re(s) > 0$
ramp	$t \cdot u(t)$	$\dfrac{1}{s^2}$	$Re(s) > 0$
nth power (for integer n)	$t^n \cdot u(t)$	$\dfrac{n!}{s^{n+1}}$	$Re(s) > 0$ $(n > -1)$

qth power (for complex q)	$t^q \cdot u(t)$	$\dfrac{\Gamma(q+1)}{s^{q+1}}$	$\mathrm{Re}(s) > 0$ $\mathrm{Re}(q) > -1$
nth root	$\sqrt[n]{t} \cdot u(t)$	$\dfrac{1}{s^{\frac{1}{n}+1}} \Gamma\!\left(\dfrac{1}{n}+1\right)$	$\mathrm{Re}(s) > 0$
nth power with frequency shift	$t^n e^{-\alpha t} \cdot u(t)$	$\dfrac{n!}{(s+\alpha)^{n+1}}$	$\mathrm{Re}(s) > -\alpha$
delayed nth power with frequency shift	$(t-\tau)^n e^{-\alpha(t-\tau)} \cdot u(t-\tau)$	$\dfrac{n! \cdot e^{-\tau s}}{(s+\alpha)^{n+1}}$	$\mathrm{Re}(s) > -\alpha$
exponential decay	$e^{-\alpha t} \cdot u(t)$	$\dfrac{1}{s+\alpha}$	$\mathrm{Re}(s) > -\alpha$
two-sided exponential decay (only for bilateral transform)	$e^{-\alpha|t|}$	$\dfrac{2\alpha}{\alpha^2 - s^2}$	$-\alpha < \mathrm{Re}(s) < \alpha$
exponential approach	$(1 - e^{-\alpha t}) \cdot u(t)$	$\dfrac{\alpha}{s(s+\alpha)}$	$\mathrm{Re}(s) > 0$
sine	$\sin(\omega t) \cdot u(t)$	$\dfrac{\omega}{s^2 + \omega^2}$	$\mathrm{Re}(s) > 0$
cosine	$\cos(\omega t) \cdot u(t)$	$\dfrac{s}{s^2 + \omega^2}$	$\mathrm{Re}(s) > 0$
hyperbolic sine	$\sinh(\alpha t) \cdot u(t)$	$\dfrac{\alpha}{s^2 - \alpha^2}$	$\mathrm{Re}(s) > |\alpha|$
hyperbolic cosine	$\cosh(\alpha t) \cdot u(t)$	$\dfrac{s}{s^2 - \alpha^2}$	$\mathrm{Re}(s) > |\alpha|$
exponentially decaying sine wave	$e^{-\alpha t} \sin(\omega t) \cdot u(t)$	$\dfrac{\omega}{(s+\alpha)^2 + \omega^2}$	$\mathrm{Re}(s) > -\alpha$
exponentially decaying cosine wave	$e^{-\alpha t} \cos(\omega t) \cdot u(t)$	$\dfrac{s+\alpha}{(s+\alpha)^2 + \omega^2}$	$\mathrm{Re}(s) > -\alpha$
natural logarithm	$\ln(t) \cdot u(t)$	$-\dfrac{1}{s}\big[\ln(s) + \gamma\big]$	$\mathrm{Re}(s) > 0$

Bessel function of the first kind, of order n	$J_n(\omega t) \cdot u(t)$	$\dfrac{\left(\sqrt{s^2 + \omega^2} - s\right)^n}{\omega^n \sqrt{s^2 + \omega^2}}$	$\text{Re}(s) > 0$ $(n > -1)$
Error function	$\text{erf}(t) \cdot u(t)$	$\dfrac{1}{s} e^{(1/4)s^2}\left(1 - \text{erf}\dfrac{s}{2}\right)$	$\text{Re}(s) > 0$

Explanatory Notes

- u(t) represents the Heaviside step function.

- δ represents the Dirac delta function.

- Γ(z) represents the Gamma function.

- γ is the Euler–Mascheroni constant.

- t, a real number, typically represents time, although it can represent any independent dimension.

- s is the complex frequency domain parameter, and Re(s) is its real part.

- α, β, τ, and ω are real numbers.

- n is an integer.

s-domain Equivalent Circuits and Impedances

The Laplace transform is often used in circuit analysis, and simple conversions to the s-domain of circuit elements can be made. Circuit elements can be transformed into impedances, very similar to phasor impedances.

Here is a summary of equivalents:

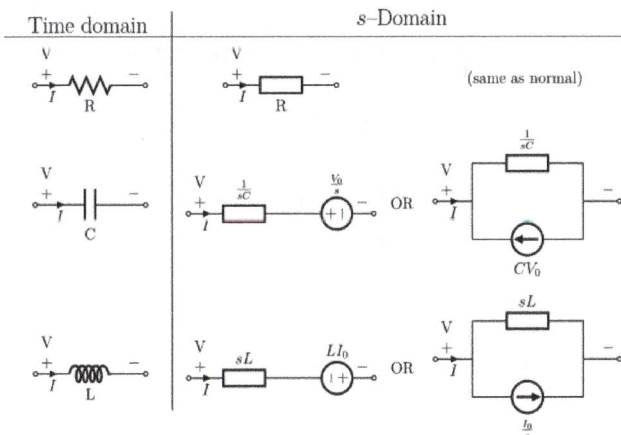

Note that the resistor is exactly the same in the time domain and the s-domain. The sources are put in if there are initial conditions on the circuit elements. For example, if a capacitor has an initial voltage across it, or if the inductor has an initial current through it, the sources inserted in the s-domain account for that.

The equivalents for current and voltage sources are simply derived from the transformations in the table above.

Examples: How to Apply the Properties and Theorems

The Laplace transform is used frequently in engineering and physics; the output of a linear time-invariant system can be calculated by convolving its unit impulse response with the input signal. Performing this calculation in Laplace space turns the convolution into a multiplication; the latter being easier to solve because of its algebraic form.

The Laplace transform can also be used to solve differential equations and is used extensively in electrical engineering. The Laplace transform reduces a linear differential equation to an algebraic equation, which can then be solved by the formal rules of algebra. The original differential equation can then be solved by applying the inverse Laplace transform. The English electrical engineer Oliver Heaviside first proposed a similar scheme, although without using the Laplace transform; and the resulting operational calculus is credited as the Heaviside calculus.

Example 1: Solving a Differential Equation

In nuclear physics, the following fundamental relationship governs radioactive decay: the number of radioactive atoms N in a sample of a radioactive isotope decays at a rate proportional to N. This leads to the first order linear differential equation

$$\frac{dN}{dt} = -\lambda N,$$

where λ is the decay constant. The Laplace transform can be used to solve this equation.

Rearranging the equation to one side, we have

$$\frac{dN}{dt} + \lambda N = 0.$$

Next, we take the Laplace transform of both sides of the equation:

$$\left(s\tilde{N}(s) - N_0 \right) + \lambda \tilde{N}(s) = 0,$$

where

$$\tilde{N}(s) = \mathcal{L}\{N(t)\}$$

and

$$N_0 = N(0).$$

Solving, we find

$$\tilde{N}(s) = \frac{N_0}{s+\lambda}.$$

Finally, we take the inverse Laplace transform to find the general solution

$$N(t) = \mathcal{L}^{-1}\{\tilde{N}(s)\} = \mathcal{L}^{-1}\left\{\frac{N_0}{s+\lambda}\right\} = N_0 e^{-\lambda t},$$

which is indeed the correct form for radioactive decay.

Example 2: Deriving the Complex Impedance for a Capacitor

In the theory of electrical circuits, the current flow in a capacitor is proportional to the capacitance and rate of change in the electrical potential (in SI units). Symbolically, this is expressed by the differential equation

$$i = C\frac{dv}{dt},$$

where C is the capacitance (in farads) of the capacitor, $i = i(t)$ is the electric current (in amperes) through the capacitor as a function of time, and $v = v(t)$ is the voltage (in volts) across the terminals of the capacitor, also as a function of time.

Taking the Laplace transform of this equation, we obtain

$$I(s) = C(sV(s) - V_0),$$

where

$$I(s) = \mathcal{L}\{i(t)\},$$
$$V(s) = \mathcal{L}\{v(t)\},$$

and

$$V_0 = v(t)\big|_{t=0}.$$

Solving for $V(s)$ we have

$$V(s) = \frac{I(s)}{sC} + \frac{V_0}{s}.$$

The definition of the complex impedance Z (in ohms) is the ratio of the complex voltage V divided by the complex current I while holding the initial state V_0 at zero:

$$Z(s) = \left.\frac{V(s)}{I(s)}\right|_{V_0=0}.$$

Using this definition and the previous equation, we find:

$$Z(s) = \frac{1}{sC},$$

which is the correct expression for the complex impedance of a capacitor.

Example 3: Method of Partial Fraction Expansion

Consider a linear time-invariant system with transfer function

$$H(s) = \frac{1}{(s+\alpha)(s+\beta)}.$$

The impulse response is simply the inverse Laplace transform of this transfer function:

$$h(t) = \mathcal{L}^{-1}\{H(s)\}.$$

To evaluate this inverse transform, we begin by expanding $H(s)$ using the method of partial fraction expansion,

$$\frac{1}{(s+\alpha)(s+\beta)} = \frac{P}{s+\alpha} + \frac{R}{s+\beta}.$$

The unknown constants P and R are the residues located at the corresponding poles of the transfer function. Each residue represents the relative contribution of that singularity to the transfer function's overall shape.

By the residue theorem, the inverse Laplace transform depends only upon the poles and their residues. To find the residue P, we multiply both sides of the equation by $s + \alpha$ to get

$$\frac{1}{s+\beta} = P + \frac{R(s+\alpha)}{s+\beta}.$$

Then by letting $s = -\alpha$, the contribution from R vanishes and all that is left is

$$P = \frac{1}{s+\beta}\bigg|_{s=-\alpha} = \frac{1}{\beta-\alpha}.$$

Similarly, the residue R is given by

$$R = \frac{1}{s+\alpha}\bigg|_{s=-\beta} = \frac{1}{\alpha-\beta}.$$

Note that

$$R = \frac{-1}{\beta-\alpha} = -P$$

and so the substitution of R and P into the expanded expression for $H(s)$ gives

$$H(s) = \left(\frac{1}{\beta-\alpha}\right) \cdot \left(\frac{1}{s+\alpha} - \frac{1}{s+\beta}\right).$$

Finally, using the linearity property and the known transform for exponential decay, we can take the inverse Laplace transform of $H(s)$ to obtain

$$h(t) = \mathcal{L}^{-1}\{H(s)\} = \frac{1}{\beta-\alpha}\left(e^{-\alpha t} - e^{-\beta t}\right),$$

which is the impulse response of the system.

Example 3.1: Convolution

The same result can be achieved using the convolution property as if the system is a series of filters with transfer functions of $1/(s + a)$ and $1/(s + b)$. That is, the inverse of

$$H(s) = \frac{1}{(s+a)(s+b)} = \frac{1}{s+a} \cdot \frac{1}{s+b}$$

is

$$\mathcal{L}^{-1}\left\{\frac{1}{s+a}\right\} * \mathcal{L}^{-1}\left\{\frac{1}{s+b}\right\} = e^{-at} * e^{-bt} = \int_0^t e^{-ax} e^{-b(t-x)} \, dx = \frac{e^{-at} - e^{-bt}}{b-a}.$$

Example 4: Mixing Sines, Cosines, and Exponentials

Time function	Laplace transform
$e^{-\alpha t}\left[\cos(\omega t)+\left(\dfrac{\beta-\alpha}{\omega}\right)\sin(\omega t)\right]u(t)$	$\dfrac{s+\beta}{(s+\alpha)^2+\omega^2}$

Starting with the Laplace transform

$$X(s)=\frac{s+\beta}{(s+\alpha)^2+\omega^2},$$

we find the inverse transform by first adding and subtracting the same constant α to the numerator:

$$X(s)=\frac{s+\alpha}{(s+\alpha)^2+\omega^2}+\frac{\beta-\alpha}{(s+\alpha)^2+\omega^2}.$$

By the shift-in-frequency property, we have

$$x(t)=e^{-\alpha t}\mathcal{L}^{-1}\left\{\frac{s}{s^2+\omega^2}+\frac{\beta-\alpha}{s^2+\omega^2}\right\}$$

$$=e^{-\alpha t}\mathcal{L}^{-1}\left\{\frac{s}{s^2+\omega^2}+\left(\frac{\beta-\alpha}{\omega}\right)\left(\frac{\omega}{s^2+\omega^2}\right)\right\}$$

$$=e^{-\alpha t}\left[\mathcal{L}^{-1}\left\{\frac{s}{s^2+\omega^2}\right\}+\left(\frac{\beta-\alpha}{\omega}\right)\mathcal{L}^{-1}\left\{\frac{\omega}{s^2+\omega^2}\right\}\right]$$

Finally, using the Laplace transforms for sine and cosine (see the table, above), we have

$$x(t)=e^{-\alpha t}\left[\cos(\omega t)u(t)+\left(\frac{\beta-\alpha}{\omega}\right)\sin(\omega t)u(t)\right].$$

$$x(t)=e^{-\alpha t}\left[\cos(\omega t)+\left(\frac{\beta-\alpha}{\omega}\right)\sin(\omega t)\right]u(t).$$

Example 5: Phase Delay

Time function	Laplace transform
$\sin(\omega t+\varphi)$	$\dfrac{s\sin(\varphi)+\omega\cos(\varphi)}{s^2+\omega^2}$
$\cos(\omega t+\varphi)$	$\dfrac{s\cos(\varphi)-\omega\sin(\varphi)}{s^2+\omega^2}.$

Starting with the Laplace transform,

$$X(s) = \frac{s\sin(\varphi) + \omega\cos(\varphi)}{s^2 + \omega^2}$$

we find the inverse by first rearranging terms in the fraction:

$$X(s) = \frac{s\sin(\varphi)}{s^2 + \omega^2} + \frac{\omega\cos(\varphi)}{s^2 + \omega^2} = \sin(\varphi)\left(\frac{s}{s^2 + \omega^2}\right) + \cos(\varphi)\left(\frac{\omega}{s^2 + \omega^2}\right).$$

We are now able to take the inverse Laplace transform of our terms:

$$x(t) = \sin(\varphi)\mathcal{L}^{-1}\left\{\frac{s}{s^2 + \omega^2}\right\} + \cos(\varphi)\mathcal{L}^{-1}\left\{\frac{\omega}{s^2 + \omega^2}\right\}$$

$$= \sin(\varphi)\cos(\omega t) + \sin(\omega t)\cos(\varphi).$$

This is just the sine of the sum of the arguments, yielding:

$$x(t) = \sin(\omega t + \varphi).$$

We can apply similar logic to find that

$$\mathcal{L}^{-1}\left\{\frac{s\cos\varphi - \omega\sin\varphi}{s^2 + \omega^2}\right\} = \cos(\omega t + \varphi).$$

Example 6: Determining Structure of Astronomical Object from Spectrum

The wide and general applicability of the Laplace transform and its inverse is illustrated by an application in astronomy which provides some information on the *spatial distribution* of matter of an astronomical source of radio-frequency thermal radiation too distant to resolve as more than a point, given its flux density spectrum, rather than relating the *time* domain with the spectrum (frequency domain).

Assuming certain properties of the object, e.g. spherical shape and constant temperature, calculations based on carrying out an inverse Laplace transformation on the spectrum of the object can produce the only possible model of the distribution of matter in it (density as a function of distance from the center) consistent with the spectrum. When independent information on the structure of an object is available, the inverse Laplace transform method has been found to be in good agreement.

The equation of motion of the system excited by an arbitrary force F(t) is

$$m\ddot{x} + c\dot{x} + kx = F(t)$$

Taking its Laplace transformation, we find

$$m\left[s^2\bar{x}(s)-s\dot{x}(0)-\dot{x}(0)\right]+c\left[s\bar{x}(s)-x(0)\right]+k\bar{x}(s)=\bar{F}(s)$$

Solving for $\bar{x}(s)$ we obtain the equation

$$\bar{x}(s)=\frac{\bar{F}(s)}{ms^2+cs+k}+\frac{(ms+c)x(0)+m\dot{x}(0)}{ms^2+cs+k}$$

The response x(t) is found from the inverse Laplace transform of the above equation, the first term represent the forced vibration and second term represents free vibration due to initial conditions.

For the more general case, the above equation can be written in the form

$$\bar{x}(s)=\frac{A(s)}{B(s)}$$

Where A(s) and B(s) are polynomials and B(s), in general, is of higher-order than A(s).

If only the forced solution is considered, we can define the impedance transform as

$$\frac{\bar{F}(s)}{\bar{x}(s)}=Z(s)=ms^2+cs+k$$

Its reciprocal is the admittance transform

$$G(s)=\frac{1}{Z(s)}=\frac{1}{ms^2+cs+k}=\frac{1}{m\left(s^2+2\zeta\omega_n s+\omega_n^2\right)}$$

$$\bar{F}(s)(imput)\rightarrow H(s)\rightarrow\bar{x}(s)(output)$$

Ex. Drive the impulse response of SDOF by Laplace Transform.

$$G(s)=\frac{1}{m\left(s^2+2\zeta\omega_n+\omega_n\right)}$$

$$=\frac{1}{2i\omega_n m}\left(\frac{1}{s+\zeta\omega_n+i\omega_d}-\frac{1}{s+\zeta\omega_n+i\omega_d}\right)$$

But, in general

$$L^{-1}\frac{1}{s-\alpha}=e^{\alpha t}$$

Hence, we get impulse response

$$g(t) = L^{-1} G(s) = L^{-1} \left(\frac{1}{s + \zeta \omega_n - i\omega_d} - \frac{1}{s + \zeta \omega_n + i\omega_d} \right)$$

$$= \frac{1}{2i\omega_d m} \left[e^{-(\zeta \omega_n - i\omega_d)t} - e^{-(\zeta \omega_n + i\omega_d)t} \right]$$

$$= \frac{1}{m\omega_d} e^{-\zeta \omega_n t} \sin \omega_d t$$

Which is precisely same as given earlier.

Ex. Determine the step response of a damped single-degree-of-freedom system by the Laplace transform method.

Laplace transform of step load can be written as $\dfrac{1}{s}$

$$\therefore \delta(t) = L^{-1} \frac{G(s)}{s} = L^{-1} \frac{1}{ms \left(s^2 + 2\zeta \omega_n s + \omega_n^2 \right)}$$

$$= \frac{1}{m\omega_n^2} L^{-1} \left(\frac{1}{s} - \frac{\zeta \omega_n + i\omega_d}{2i\omega_d} \cdot \frac{1}{s + \zeta \omega_n - i\omega_d} + \frac{\zeta \omega_n - i\omega_d}{2i\omega_d} \cdot \frac{1}{s + \zeta \omega_n - i\omega_d} \right)$$

Therefore, the step response can be obtained as follows:

$$\delta(t) = \frac{1}{k} \left[1 - \frac{\zeta \omega_n + i\omega_d}{2i\omega_d} e^{-(\zeta \omega_n - i\omega_d)t} + \frac{\zeta \omega_n - i\omega_d}{2i\omega_d} e^{-(\zeta \omega_n + i\omega_d)t} \right]$$

$$= \frac{1}{k} \left[1 - \frac{1}{(1-\zeta^2)^{\frac{1}{2}}} e^{-\zeta \omega_n t} \cos(\omega_d t - \psi) \right]$$

When $\psi = \tan^{-1} \dfrac{\zeta}{(1-\zeta^2)^{\frac{1}{2}}}$

4

Numerical Analysis of Dynamic System

The differential equation of motion of a dynamic system may be linear or non-linear. These vary as systems with different degrees of freedom (DOF) are taken into consideration. Aircraft engineering is best understood in confluence with the major topics listed in the following chapter.

Numerical Solution of a Dynamic System

The differential equation of motion for a dynamic system, which may be linear or non-linear, can be expressed as

$$\ddot{x} = f(x, \dot{x}, t)$$

$$\left.\begin{array}{l} x_1 = x(0) \\ \dot{x}_1 = \dot{x}(0) \end{array}\right\} \text{initial condition}$$

There are two methods to solve the second-order differential equation:

1) Method one – Integrate the second-order differential equation without change in form.

2) Method two – the second-order differential equation is reduced to two first order equation before integration. The equation then takes the form:

$$\dot{x} = y$$
$$\dot{y} = \ddot{x} = f(x, y, z)$$

Method I

The Taylor series expansion can be used to derive finite differential scheme.

$$x(t + \Delta t) = x(t) + \Delta t\, \dot{x}(t) + \frac{(\Delta t)^2}{2!}\ddot{x}(t) + \frac{\Delta t^3}{3!}\dddot{x}(t) + \ldots\ldots \rightarrow \quad (1)$$

$$x(t - \Delta t) = x(t) - \Delta t\, \dot{x}(t) + \frac{(\Delta t)^2}{2!}\ddot{x}(t) - \frac{\Delta t^3}{3!}\dddot{x}(t) + \ldots\ldots \rightarrow \quad (2)$$

Subtracting (2) from (1) and ignoring higher-order terms we obtain

$$\dot{x} = \frac{1}{2\Delta t}\{x(1+\Delta t) - x(t+\Delta t)\} \rightarrow \qquad (3)$$

Similarly, adding equations (1) and (2) neglecting higher-order terms

$$\ddot{x}(t) = \frac{1}{\Delta t^2}\{x(t-\Delta t) - 2x(t) + x(t+\Delta t)\} \rightarrow \qquad (4)$$

The displacement solution for time $t + \Delta t$ is obtained by considering at time i.e.,

$$M\ddot{x}(t) + c\dot{x}(t) + Kx(t) = R(t)$$

Substituting the relation for $\ddot{x}(t)$ and $\dot{x}(t)$ and rearranging we obtain

$$\left|\left(\frac{1}{\Delta t^2}M + \frac{1}{2\Delta t}C\right)x(t+\Delta t) = R(t) - \left(K - \frac{2}{\Delta t^2}M\right)x(t) - \left(\frac{1}{\Delta t^2}M - \frac{1}{2\Delta t}C\right)x(t-\Delta t)\right|$$

From which we can solve for $x(t+\Delta t)$. It should be noted that the solution of $x(t+\Delta t)$ is thus based on using equilibrium equation at time t. For this reason the integration procedure is called on explicit integration method, and it is noted that such integration schemes do not require a factorization of the (effective) stiffness matrix in the step-by-step solution.

To solve we need $x(t)$, $x(t+\Delta t)$ and $x(t-\Delta t)$ values. Using equation (3) and (4) we have

$$\Delta t^2 \ddot{x}(t) = x(t-\Delta t) - 2x(t) + x(t+\Delta t) -----(5)$$

$$2\Delta t \dot{x}(t) = -x(t-\Delta t) + x(t+\Delta t) ------(6)$$

Subtracting equation (6) from equation (5) we get

$$\Rightarrow \Delta t^2 \ddot{x}(t) - 2\Delta \dot{x}(t) = 2x(t-\Delta t) - 2x(t)$$

$$\text{or, } \quad x(t-\Delta t) = x(t) - \Delta t\dot{x}(t) + \frac{\Delta t^2}{2}\ddot{x}(t) ---- (7)$$

Step-by-step Procedure to use Difference Method

A.   Initial Calculations:

1. From stiffness matrix K, mass matrix M, and damping matrix C.

2. Initialize $U(0), \dot{U}(0) \, and \, \ddot{U}(0)$

3. Select time step cr $\Delta t, \Delta t < \Delta t_{cr}$ and calculate integration constants.

$$a_0 = \frac{1}{\Delta t^2}; \; a_1 = \frac{1}{2\Delta t}; a_2 = 2a_0 \,; a_3 = \frac{1}{a_2}$$

4. Calculate $U(-\Delta t) = U(0) - \Delta t \dot{U}(0) + a_3 \ddot{U}(0)$

5. From effective mass matrix, $\hat{M} = a_0 \, M + a_1 \, C$

6. Triangularize, $\hat{M} : \hat{M} = LDL^T$

B.   For each time step calculation

1. Calculate effective load at time t :

$$\hat{R}(t) = R(t) - (K - a_2 \, M)U(t) - (a_0 M - a_1 C)U(t - \Delta t)$$

2. Solve for displacements at time $t + \Delta t$

$$LDL^T \cdot u(t + \Delta t) = R(t)$$

3. If required, calculate acceleration and velocity at time t

$$\ddot{U}(t) = a_0 \{ U(t - \Delta t) - 2U(t) + U(t + \Delta t) \}$$

$$\dot{U}(t) = a_1 \{ U(t + \Delta t) - U(t - \Delta t) \}$$

Solution of Δt value:

Δt is smaller than a critical value, Δt_{cr}, which can be calculated from the mass and stiffness properties of the complete assemblage. More, specifically, we will show that to obtain a valid solution

$$\Delta t \le \Delta t_{cr} = \frac{T_n}{\pi}$$

Where T_n is the smallest period of the finite element assemblage with n degrees of freedom.

Ex. Consider simple system for which the governing equilibrium equations are;

$$\begin{bmatrix} 2 & 0 \\ 0 & 1 \end{bmatrix} \begin{Bmatrix} \ddot{u}_1 \\ \ddot{u}_2 \end{Bmatrix} + \begin{bmatrix} 6 & -2 \\ -2 & 4 \end{bmatrix} \begin{Bmatrix} u_1 \\ u_2 \end{Bmatrix} = \begin{Bmatrix} 0 \\ 10 \end{Bmatrix}$$

Time period for 1st two modes are:

$$T_1 = 4.45\,s, T_2 = 2.8\,s.$$

$$\therefore \Delta t = \frac{T_2}{10} = \frac{2.8}{10} = 0.28\,s.$$

Calculate the response of the system assuming $^0U = 0,\ ^0\dot{U} = 0$

The first step :

$$\begin{bmatrix} 2 & 0 \\ 0 & 1 \end{bmatrix} \{ ^0\ddot{U} \} + \begin{bmatrix} 6 & -2 \\ -2 & 4 \end{bmatrix} \begin{Bmatrix} 0 \\ 0 \end{Bmatrix} = \begin{Bmatrix} 0 \\ 10 \end{Bmatrix}$$

$$\Rightarrow \{ ^0\ddot{U} \} = \begin{Bmatrix} 0 \\ 10 \end{Bmatrix}$$

Now consider $\Delta t = 0.28\,s$

$$a_0 = \frac{1}{(0.28)^2} = 12.8 \quad a_1 = \frac{1}{2 \times 0.28} = 1.79$$

$$a_2 = 2a_0 = 25.5 \quad a_3 = \frac{1}{a_3} = 0.0392$$

Hence,

Hence, $^{-\Delta t}U = \begin{Bmatrix} 0 \\ 0 \end{Bmatrix} - 0.28 \begin{Bmatrix} 0 \\ 0 \end{Bmatrix} + 0.0392 \begin{Bmatrix} 0 \\ 10 \end{Bmatrix} = \begin{Bmatrix} 0 \\ 0.392 \end{Bmatrix}$

$\hat{M} = 12.8 \begin{bmatrix} 2 & 0 \\ 0 & 1 \end{bmatrix} + 1.79 \begin{bmatrix} 0 & 0 \\ 0 & 0 \end{bmatrix} = \begin{bmatrix} 25.5 & 0 \\ 0 & 12.8 \end{bmatrix}$

Hence, we need to solve the following equations for each time step;

$$\begin{bmatrix} 25.5 & 0 \\ 0 & 12.8 \end{bmatrix} \{ ^{t+\Delta t}U \} = {}^t\{R\}$$

Calculating the solution for each time step, we obtain

Time	Δt	$2\Delta t$	$3\Delta t$	$4\Delta t$	$5\Delta t$	$6\Delta t$	$7\Delta t$	$8\Delta t$	$9\Delta t$	10Δ	11Δ	12Δ
^tU	0	0.030	0.16	0.48	1.0	1.7	2.4	2.9	3.0	2.77	2.0	1.0
	0.39	1.45	2.83	4.14	5.0	5.2	4.9	4.1	3.3	2.7	2.5	2.6

Other Implicit Integration Methods are Houbolt Method and Wilson Method.

The Newmark integration scheme can be understood to be an extension of the linear acceleration method.

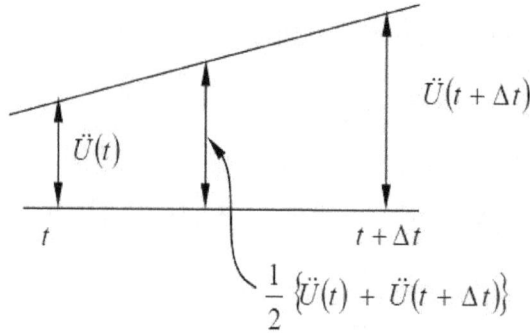

$$\frac{1}{2}\left\{\ddot{U}(t) + \ddot{U}(t + \Delta t)\right\}$$

The following assumption are used

$$\dot{U}(t + \Delta t) = \dot{U}(t) + \left[(1 - \delta)\ddot{U}(t) + \delta\ddot{U}(t + \Delta t)\right]\Delta t$$

$$U(t + \Delta t) = U(t) + \Delta t\dot{U}(t) + \left[\left(\frac{1}{2} - \alpha\right)\ddot{U}(t) + \alpha\ddot{U}(t + \Delta t)\right]\Delta t^2$$

Where $\delta = \frac{1}{2}$ and $\alpha = \frac{1}{4}$ originally proposed by Newmark as an unconditionally stable scheme.

A.   Initial calculations:

1. Form Stiffness K, Mass M and Damping C matrices.

2. Initialize $U(0), \dot{U}(0) \, and \, \ddot{U}(0)$

3. Select time step size Δt , parameters α and δ and calculate integration constants.

$$\delta \geq 0.5, \ \alpha \geq 0.25(0.5 + \delta)$$

$$a_0 = \frac{1}{\alpha \Delta t^2}; \ a_1 = \frac{\delta}{\alpha \Delta t}; \ a_2 = \frac{1}{\alpha \Delta t}; \ a_3 = \frac{1}{2\alpha} - 1$$

$$a_4 = \frac{\delta}{\alpha}; \ a_5 = \frac{\Delta t}{2}\left(\frac{\delta}{\alpha} - 2\right); \ a_6 = \Delta t(1 - \delta); \ a_7 = \delta \Delta t$$

4. Form effective stiffness matrix $\hat{K}$: $\hat{K} = K + a_0 M + a_1 C$

5. Triangularize $\hat{K}$: $\hat{K} = LDL^T$

B. For each time step

1. Calculate effective loads at time $(t + \Delta t)$

$$R(t + \Delta t) = R(t + \Delta t) + M\left\{a_0 U(t) + a_2 \dot{U}(t) + a_3 \ddot{U}(t)\right\}$$

$$+ C\left\{a_1 U(t) + a_5 \ddot{U}(t)\right\}$$

2. Solve for displacement at $t + \Delta t$

$$LDL^T \cdot U(t + \Delta t) = \hat{R}(t + \Delta t)$$

3. Calculation of accelerations and velocities at $t + \Delta t$

$$\ddot{U}(t + \Delta t) = a_0 \left\{U(t + \Delta t) - U(t)\right\} - a_2 \dot{U}(t) - a_3 \ddot{U}(t)$$
$$\dot{U}(t + \Delta t) = \dot{U}(t) + a_6 \ddot{U}(t) + a_7 \ddot{U}(t + \Delta t)$$

$\therefore$ Earlier example by Newmarks method

$$\Delta t = 0.28\,s. \quad \alpha = 0.25 \quad \delta = 0.5$$

$$U_0 = \begin{Bmatrix} 0 \\ 0 \end{Bmatrix}, \ \dot{U}_0 = \begin{Bmatrix} 0 \\ 0 \end{Bmatrix}, \ \ddot{U}_0 = \begin{Bmatrix} 0 \\ 10 \end{Bmatrix}$$

Integration Constants are

$$a_0 = 51.0 \quad a_1 = 7.14 \quad a_2 = 14.3 \quad a_3 = 1.0$$
$$a_4 = 1.0 \quad a_5 = 0.0 \quad a_6 = 0.14 \quad a_7 = 0.14$$

Thus the effective stiffness matrix is

$$\hat{K} = \begin{bmatrix} 6 & -2 \\ -2 & 4 \end{bmatrix} + 51.0 \begin{bmatrix} 2 & 0 \\ 0 & 1 \end{bmatrix} = \begin{bmatrix} 108 & -2 \\ -2 & 55 \end{bmatrix}$$

For each time step we need to evaluate

$$\hat{R}(t + \Delta t) = \begin{bmatrix} 0 \\ 10 \end{bmatrix} + \begin{bmatrix} 2 & 0 \\ 0 & 1 \end{bmatrix} \left(51 U(t) + 14.3 \dot{U}(t) + 1.0 \ddot{U}(t)\right)$$

Then, $\hat{K}U(t+\Delta t) = \hat{R}(t+\Delta t)$

And $\ddot{U}(t+\Delta t)=51.0\big(U(t+\Delta t)-U(t)-14.3\dot{U}(t)+1.0\ddot{U}(t)\big)$

$$\dot{U}(t+\Delta t)=\dot{U}(t)+0.14\ddot{U}(t)+0.14\ddot{U}(t+\Delta t)$$

Performing the above calculation, we obtain

Tim	Δt	$2\Delta t$	$3\Delta t$	$4\Delta t$	$5\Delta t$	6Δ	7Δ	$8\Delta t$	9Δ	10Δ	11Δ	12Δ
U	0.006	0.050	0.18	0.48	0.96	1.5	2.2	2.7	3.0	2.8	2.2	1.4
	0.364	1.35	2.68	4.0	4.95	5.3	5.1	4.4	3.6	2.9	2.4	2.3

Method II (Runge Kutta Method)

In Runga Kutta method the second order differential equation is first reduced to two first order equations. As an example, consider the differential equation for the SDOF system, which may be written as

$$\ddot{x}=\frac{1}{m}\big[F(t)-Kx-C\dot{x}\big]=f(x,\dot{x},t)$$

By letting $\dot{x}=y$ the above equation is reduced to the following two first order equations:

$$\dot{x}=y$$

$$\dot{y}=\ddot{x}=f(x,y,t)$$

Both x and y in the neighborhood of x_i and y_i can be expressed in terms of the Taylor series.

Letting the time increment be $h=\Delta t$, we have

$$x = x_i +\left(\frac{dx}{dt}\right)_i h +\left(\frac{d^2x}{dt^2}\right)_i \frac{h^2}{2} + ----$$

$$y = y_i +\left(\frac{dy}{dt}\right)_i h +\left(\frac{d^2y}{dt^2}\right)_i \frac{h^2}{2} + ----$$

Ignoring higher-order derivative, we can write

$$x = x_i +\left(\frac{dx}{dt}\right)_{iav} .h$$

$$y = y_i + \left(\frac{dy}{dt}\right)_{iav} .h$$

If we use Simpson's rule, the average slope in the interval h becomes, i.e.

$$\left(\frac{dy}{dt}\right)_{iav} = \frac{1}{6}\left[\left(\frac{dy}{dt}\right)_{t_i} + 4\left(\frac{dy}{dt}\right)_{t_i + h/2} + \left(\frac{dy}{dt}\right)_{t_i + h}\right]$$

The Runge-Kutta is very similar to the preceding computations, except that center terms of the given equation is split into two terms and four values of x, y and f are computed for each point i as follows:

t	x	$y = \dot{x}$	$f = \dot{y} = \ddot{x}$
$T_1 = t_i$	$X_1 = x_i$	$Y_1 = y_i$	$F_1 = f(T_1, X_1, Y_1)$
$T_2 = t_i + \dfrac{h}{2}$	$X_2 = x_i + Y_1 \dfrac{h}{2}$	$Y_2 = y_i F_1 \dfrac{h}{2}$	$F_2 = f(T_2, X_2, Y_2)$
$T_3 = t_i + \dfrac{h}{2}$	$X_3 = x_i + Y_2 \dfrac{h}{2}$	$Y_3 = y_i F_2 \dfrac{h}{2}$	$F_3 = f(T_3, X_3, Y_3)$
$T_4 = t_i + h$	$X_4 = x_i + Y_3 h$	$Y_4 = y_i F_3 \dfrac{h}{2}$	$F_4 = f(T_4, X_4, Y_4)$

These equations are then used in the following recurrence formula:

$$x_{i+1} = x_i + \frac{h}{6}[Y_1 + 2Y_2 + 2Y_3 + Y_4]$$

$$y_{i+1} = y_i + \frac{h}{6}[F_1 + 2F_2 + 2F_3 + F_4]$$

Where it is recognized that the four values of Y divided by 6 represent an average slope dx/dt and the four value of F divided by 6 results in an average of dy/dt as define by eq, (arrange acceleration).

Solve numerically the differential equation

$$4\ddot{x} + 2000x = F(t)$$

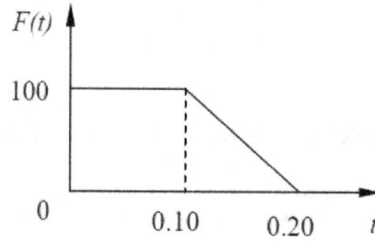

With initial condition $x_1 = \dot{x}_1 = 0$

Solution: The differential equation of motion is

$$\ddot{x}_1 = \frac{1}{4}F(t) - 500x$$

Let $y = \dot{x}$ then

$$\dot{y} = \ddot{x} = \frac{1}{4}F(t) - 500$$

$$\omega = \sqrt{\frac{2000}{4}} = \sqrt{500} = 10\sqrt{5} = 22.36 \; rad/s$$

$$T = \frac{2\pi}{\omega} = 0.281s$$

$$\Delta t \leq \frac{T}{10} \leq \frac{0.281}{10} \leq 0.0281 \approx 0.02$$

	t	x	$y = \dot{x}$	f
$t_1 =$	0	0	0	25
	0.01	0	0.25	25
	0.01	0.0025	0.25	23.75
$t_2 =$	0.02	.0050	0.475	22.50

Calculate

$$x_2 = 0 + \frac{0.02}{6}[0 + .0.50 + 0.50 + 0.475] = 0.00491667$$

$$y_2 = 0 + \frac{0.02}{6}[25 + 50 + 47.50 + 22.50] = 0.483333$$

To continue to point 3, we repeat the above table

$t_2 =$	0.02	0.00491667	0.4833333	22.541665
	0.03	0.0097500	0.70874997	20.12500
	0.03	0.01200417	0.6845833	18.997915
$t_3 =$	0.04	0.01860834	0.8632912	15.695830

We then calculate x_3 *and* y_3

$$x_3 = 0.00491667 + \frac{0.02}{6}\left[0.483333 + 2 \times 0.70874997 + 2 \times 0.6845833 + 0.8632912\right]$$

$$= 0.01869431$$

$$y_3 = 0.4833333 + \frac{0.02}{6}\left[22.541665 + 2 \times 20.125 + 2 \times 18.997915 + 15.695830\right]$$

$$= 0.4833333 + 0.38827775$$

The Newmark Method

Other Implicit Integration Methods are Houbolt Method and Wilson Method.

The Newmark integration scheme can be understood to be an extension of the linear acceleration method.

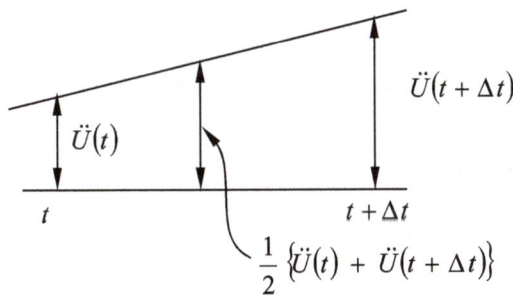

The following assumption are used

$$\dot{U}(t+\Delta t) = \dot{U}(t) + \left[(1-\delta)\ddot{U}(t) + \delta\ddot{U}(t+\Delta t)\right]\Delta t$$

$$U(t+\Delta t) = U(t) + \Delta t\dot{U}(t) + \left[\left(\frac{1}{2}-\alpha\right)\ddot{U}(t) + \alpha\ddot{U}(t+\Delta t)\right]\Delta t^2$$

Where $\delta = \dfrac{1}{2}$ and $\alpha = \dfrac{1}{4}$ originally proposed by Newmark as an unconditionally stable scheme.

A.   Initial calculations:

1. Form Stiffness K, Mass M and Damping C matrices.

2. Initialize $U(0), \dot{U}(0) \, and \, \ddot{U}(0)$

3. Select time step size Δt, parameters α and δ and calculate integration constants.

$$\delta \geq 0.5, \quad \alpha \geq 0.25(0.5 + \delta)$$

$$a_0 = \frac{1}{\alpha \Delta t^2}; \, a_1 = \frac{\delta}{\alpha \Delta t}; \, a_2 = \frac{1}{\alpha \Delta t}; \, a_3 = \frac{1}{2\alpha} - 1$$

$$a_4 = \frac{\delta}{\alpha}; \, a_5 = \frac{\Delta t}{2}\left(\frac{\delta}{\alpha} - 2\right); \, a_6 = \Delta t(1 - \delta); \, a_7 = \delta \Delta t$$

4. Form effective stiffness matrix $\hat{K}$: $\hat{K} = K + a_0 M + a_1 C$

5. Triangularize $\hat{K}$: $\hat{K} = LDL^T$

B.   For each time step

1. Calculate effective loads at time $(t + \Delta t)$

$$R(t + \Delta t) = R(t + \Delta t) + M\left\{a_0 U(t) + a_2 \dot{U}(t) + a_3 \ddot{U}(t)\right\}$$

$$+ C\left\{a_1 U(t) + a_4 \dot{U}(t) + a_5 \ddot{U}(t)\right\}$$

2. Solve for displacement at $(t + \Delta t)$

$$LDL^T . U(t + \Delta t) = \hat{R}(t + \Delta t)$$

3. Calculation of accelerations and velocities at $(t + \Delta t)$

$$\ddot{U}(t + \Delta t) = a_0 \left\{U(t + \Delta t) - U(t)\right\} - a_2 \dot{U}(t) - a_3 \ddot{U}(t)$$
$$\dot{U}(t + \Delta t) = \dot{U}(t) + a_6 \ddot{U}(t) + a_7 \ddot{U}(t + \Delta t)$$

Earlier example by Newmarks method

Δt =0.28s. α =0.25 δ =0.5

$$U_0 = \begin{Bmatrix} 0 \\ 0 \end{Bmatrix}, \quad \dot{U}_0 = \begin{Bmatrix} 0 \\ 0 \end{Bmatrix}, \quad \ddot{U}_0 = \begin{Bmatrix} 0 \\ 0 \end{Bmatrix}$$

Integration Constants are

$$a_0 = 51.0 \quad a_1 = 7.14 \quad a_2 = 14.3 \quad a = 1.0$$

$$a_4 = 1.0 \quad a_5 = 0.0 \quad a_6 = 0.14 \quad a_7 = 0.14$$

Thus the effective stiffness matrix is

$$\hat{K} = \begin{bmatrix} 6 & -2 \\ -2 & 4 \end{bmatrix} + 51.0 \begin{bmatrix} 2 & 0 \\ 0 & 1 \end{bmatrix} = \begin{bmatrix} 108 & -2 \\ -2 & 55 \end{bmatrix}$$

For each time step we need to evaluate

$$\hat{R}(t + \Delta t) = \begin{bmatrix} 0 \\ 10 \end{bmatrix} + \begin{bmatrix} 2 & 0 \\ 0 & 1 \end{bmatrix} \left(51 U(t) + 14.3 \dot{U}(t) + 1.0 \ddot{U}(t) \right)$$

Then, $\hat{K} U(t + \Delta t) = \hat{R}(t + \Delta t)$

and
$$\ddot{U}(t + \Delta t) = 51.0 \left(U(t + \Delta t) - U(t) \right) - 14.3 \dot{U}(t) + 1.0 \ddot{U}t$$
$$\dot{U}(t + \Delta t) = \dot{U}(t) + 0.14 \ddot{U}(t) + 0.14 \ddot{U}(t + \Delta t)$$

Performing the above calculation, we obtain

Time	Δt	$2\Delta t$	$3\Delta t$	$4\Delta t$	$5\Delta t$	$6\Delta t$	$7\Delta t$	$8\Delta t$	$9\Delta t$	$10\Delta t$	$11\Delta t$	$12\Delta t$
$U(t)$	0.00673	0.0505	0.189	0.485	0.961	1.58	2.23	2.76	3.0	2.85	2.28	1.40
	0.364	1.35	2.68	4.0	4.95	5.34	5.13	4.48	3.64	2.90	2.44	2.31

Method II (Runge Kutta Method)

In Runga Kutta method the second order differential equation is first reduced to two first

order equations. As an example, consider the differential equation for the SDOF system, which

may be written as

$$\ddot{x} = \frac{1}{m} \left[F(t) - Kx - C\dot{x} \right] = f(x, \dot{x}, t)$$

By letting $\dot{x} = y$, the above equation is reduced to the following two first order equations:

$$\dot{x} = y$$

$$\dot{y} = \ddot{x} = f(x, y, t)$$

Both x and y in the neighborhood of x_i and y_i can be expressed in terms of the Taylor series.

Letting the time increment be h= Δt , we have

$$x = x_i + \left(\frac{dx}{dt}\right)_i h + \left(\frac{d^2x}{dt^2}\right)_i \frac{h^2}{2} + - - - -$$

$$y = y_i + \left(\frac{dy}{dt}\right)_i h + \left(\frac{d^2y}{dt^2}\right)_i \frac{h^2}{2} + - - - -$$

Ignoring higher-order derivative, we can write

$$x = x_i + \left(\frac{dx}{dt}\right)_{iav} . h$$

$$y = y_i + \left(\frac{dy}{dt}\right)_{iav} . h$$

If we use Simpson's rule, the average slope in the interval h becomes, i.e.

$$\left(\frac{dx}{dt}\right)_{iav} = \frac{1}{6}\left[\left(\frac{dy}{dt}\right)_{t_i} + 4\left(\frac{dy}{dt}\right)_{t_i+h/2} + \left(\frac{dy}{dt}\right)_{t_i+h}\right]$$

The Runge-Kutta is very similar to the preceding computations, except that center terms of the given equation is split into two terms and four values of x, y and f are computed for each point i

as follows:

t	x	$y = \dot{x}$	$f = \dot{y} = \ddot{x}$
$T_1 = t_i$	$X_1 = x_i$	$Y_1 = y_i$	$F_1 = f(T_1, X_1, Y_1)$
$T_2 = t_i + \dfrac{h}{2}$	$X_2 = x_i + Y_1\dfrac{h}{2}$	$Y_2 = y_i\, F_1\dfrac{h}{2}$	$F_2 = f(T_2, X_2, Y_2)$
$T_3 = t_i + \dfrac{h}{2}$	$X_3 = x_i + Y_2\dfrac{h}{2}$	$Y_3 = y_i\, F_2\dfrac{h}{2}$	$F_3 = f(T_3, X_3, Y_3)$
$T_4 = t_i + h$	$X_4 = x_i + Y_3 h$	$Y_4 = y_i\, F_3\dfrac{h}{2}$	$F_4 = f(T_4, X_4, Y_4)$

These equations are then used in the following recurrence formula:

$$x_{i+1} = x_i + \frac{h}{6}\left[Y_1 + 2Y_2 + 2Y_3 + Y_4\right]$$

$$Y_{i+1} = Y_i + \frac{h}{6}\left[F_1 + 2F_2 + 2F_3 + F_4\right]$$

Where it is recognized that the four values of Y divided by 6 represent an average slope dx/dt and the four value of F divided by 6 results in an average of dy/dt as define by eq, (arrange acceleration).

Solve numerically the differential equation

$$4\ddot{x} + 2000x = F(t)$$

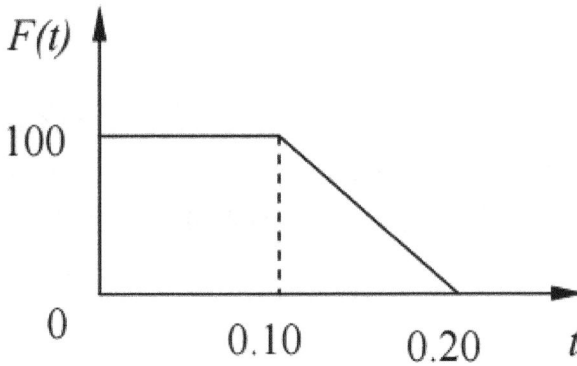

With initial condition $x_1 = \dot{x}_1 = 0$

Solution: The differential equation of motion is

$$\ddot{x}_1 \qquad = \frac{1}{4}F(t) - 500x$$

let $y = \dot{x}$ then

$$\dot{y} = \ddot{x} = \frac{1}{4}F(t) - 500$$

$$\omega = \sqrt{\frac{2000}{4}} = \sqrt{500} = 10\sqrt{5} = 22.36 \ rad/s$$

$$T = \frac{2\pi}{\omega} = 0.281s$$

$$\Delta t \le \frac{T}{10} \le \frac{0.281}{10} \le 0.0281 \approx 0.02$$

	t	x	$y = \ddot{x}$	f
$t_1 =$	0	0	0	25
	0.01	0	0.25	25
	0.01	0.0025	0.25	23.75
$t_2 =$	0.02	.0050	0.475	22.50

Calculate

$$x_2 = 0 + \frac{0.02}{6}[0 + .0.50 + 0.50 + 0.475] = 0.00491667$$

$$y_2 = 0 + \frac{0.02}{6}[25 + 50 + 47.50 + 22.50] = 0.483333$$

To continue to point 3, we repeat the above table

	t	x	y	f
$t_2 =$	0.02	0.00491667	0.4833333	22.541665
	0.03	0.0097500	0.70874997	20.12500
	0.03	0.01200417	0.6845833	18.997915
$t_3 =$	0.04	0.01860834	0.8632912	15.695830

We then calculate x_3 and y_3

$$x_3 = 0.00491667 + \frac{0.02}{6}[0.483333 + 2 \times 0.70874997 + 2 \times 0.684583 + 0.8632912]$$

$$= 0.01869431$$

$$y_3 = 0.4833333 + \frac{0.02}{6}[22.541665 + 2 \times 20.125 + 2 \times 18.997915 + 15.695830]$$

$$= 0.4833333 + 0.388827775$$

$$= 0.87161075$$

Runge–Kutta Methods

In numerical analysis, the Runge–Kutta methods are a family of implicit and explicit iterative methods, which includes the well-known routine called the Euler Method, used in temporal discretization for the approximate solutions of ordinary differential equations. These methods were developed around 1900 by the German mathematicians C. Runge and M. W. Kutta.

The Runge–Kutta Method

The most widely known member of the Runge–Kutta family is generally referred to as "RK4", "classical Runge–Kutta method" or simply as "*the* Runge–Kutta method".

Let an initial value problem be specified as follows:

$$\dot{y} = f(t, y), \quad y(t_0) = y_0.$$

Here y is an unknown function (scalar or vector) of time t, which we would like to approximate; we are told that $\dot{y}$, the rate at which y changes, is a function of t and of y itself. At the initial time t_0 the corresponding y value is y_0 . The function f and the data t_0, y_0 are given.

Now pick a step-size $h > 0$ and define

$$y_{n+1} = y_n + \tfrac{h}{6}\left(k_1 + 2k_2 + 2k_3 + k_4\right),$$
$$t_{n+1} = t_n + h$$

for n = 0, 1, 2, 3, ..., using

$$k_1 = f(t_n, y_n),$$
$$k_2 = f(t_n + \frac{h}{2}, y_n + \frac{h}{2}k_1),$$
$$k_3 = f(t_n + \frac{h}{2}, y_n + \frac{h}{2}k_2),$$
$$k_4 = f(t_n + h, y_n + hk_3).$$

(Note: the above equations have different but equivalent definitions in different texts).

Here y_{n+1} is the RK4 approximation of $y(t_{n+1})$,, and the next value (y_{n+1}) is determined by the present value (y_n) plus the weighted average of four increments, where each increment is the product of the size of the interval, h, and an estimated slope specified by function f on the right-hand side of the differential equation.

- k_1 is the increment based on the slope at the beginning of the interval, using y (Euler's method);

- k_2 is the increment based on the slope at the midpoint of the interval, using $y + \dfrac{h}{2}k_1$;

- k_3 is again the increment based on the slope at the midpoint, but now using

$y + \dfrac{h}{2}k_2$;

- k_4 is the increment based on the slope at the end of the interval, using $y + hk_3$.

In averaging the four increments, greater weight is given to the increments at the mid-point. If f is independent of y , so that the differential equation is equivalent to a simple integral, then RK4 is Simpson's rule.

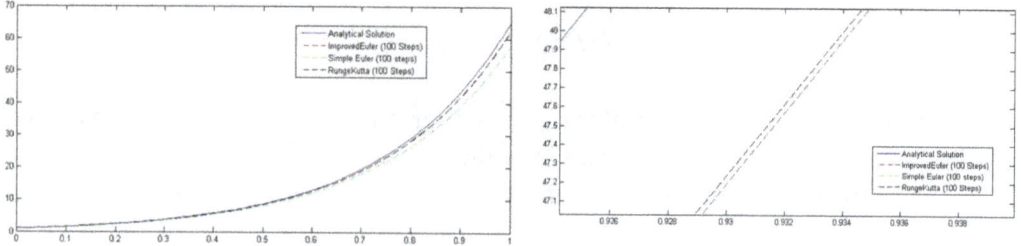

Comparison of the Runge–Kutta 4 with other lower-order methods for a given ordinary differential equation

The RK4 method is a fourth-order method, meaning that the local truncation error is on the order of $O(h^5)$, while the total accumulated error is on the order of $O(h^4)$.

Explicit Runge–Kutta Methods

The family of explicit Runge–Kutta methods is a generalization of the RK4 method mentioned above. It is given by

$$y_{n+1} = y_n + h\sum_{i=1}^{s} b_i k_i,$$

where

$$k_1 = f(t_n, y_n),$$
$$k_2 = f(t_n + c_2 h, y_n + h(a_{21}k_1)),$$
$$k_3 = f(t_n + c_3 h, y_n + h(a_{31}k_1 + a_{32}k_2)),$$
$$\vdots$$
$$k_s = f(t_n + c_s h, y_n + h(a_{s1}k_1 + a_{s2}k_2 + \cdots + a_{s,s-1}k_{s-1})).$$

(Note: the above equations have different but equivalent definitions in different texts).

To specify a particular method, one needs to provide the integer s (the number of stages), and the coefficients a_{ij} (for $1 \leq j < i \leq s$), b_i (for $i = 1, 2, ..., s$) and c_i (for $i = 2, 3, ..., s$). The matrix $[a_{ij}]$ is called the *Runge–Kutta matrix*, while the b_i and c_i are known as the *weights* and the *nodes*. These data are usually arranged in a mnemonic device, known as a *Butcher tableau* (after John C. Butcher):

$$
\begin{array}{c|ccccc}
0 & & & & & \\
c_2 & a_{21} & & & & \\
c_3 & a_{31} & a_{32} & & & \\
\vdots & \vdots & & \ddots & & \\
c_s & a_{s1} & a_{s2} & \cdots & a_{s,s-1} & \\
\hline
& b_1 & b_2 & \cdots & b_{s-1} & b_s
\end{array}
$$

The Runge–Kutta method is consistent if

$$
\sum_{j=1}^{i-1} a_{ij} = c_i \ \text{for } i = 2,\ldots,s.
$$

There are also accompanying requirements if one requires the method to have a certain order p, meaning that the local truncation error is $O(h^{p+1})$. These can be derived from the definition of the truncation error itself. For example, a two-stage method has order 2 if $b_1 + b_2 = 1$, $b_2 c_2 = 1/2$, and $a_{21} = c_2$.

In general, if an explicit s-stage Runge–Kutta method has order p, then $s \geq p$, and if $p \geq 5$, then $s > p$. The minimum s required for an explicit s-stage Runge–Kutta method to have order p is an open problem. Some values which are known are:

p	1	2	3	4	5	6	7	8
$\min s$	1	2	3	4	6	7	9	11

Examples

The RK4 method falls in this framework. Its tableau is

$$
\begin{array}{c|cccc}
0 & & & & \\
1/2 & 1/2 & & & \\
1/2 & 0 & 1/2 & & \\
1 & 0 & 0 & 1 & \\
\hline
& 1/6 & 1/3 & 1/3 & 1/6
\end{array}
$$

A slight variation of "the" Runge–Kutta method is also due to Kutta in 1901 and is called the 3/8-rule. The primary advantage this method has is that almost all of the error coefficients are smaller than in the popular method, but it requires slightly more

FLOPs (floating-point operations) per time step. Its Butcher tableau is

$$
\begin{array}{c|cccc}
0 & & & & \\
1/3 & 1/3 & & & \\
2/3 & -1/3 & 1 & & \\
1 & 1 & -1 & 1 & \\
\hline
& 1/8 & 3/8 & 3/8 & 1/8
\end{array}
$$

However, the simplest Runge–Kutta method is the (forward) Euler method, given by the formula $y_{n+1} = y_n + hf(t_n, y_n)$. This is the only consistent explicit Runge–Kutta method with one stage. The corresponding tableau is

$$
\begin{array}{c|c}
0 & \\
\hline
& 1
\end{array}
$$

Second-order Methods with Two Stages

An example of a second-order method with two stages is provided by the midpoint method:

$$
y_{n+1} = y_n + hf\left(t_n + \frac{1}{2}h, y_n + \frac{1}{2}hf(t_n, y_n)\right).
$$

The corresponding tableau is

$$
\begin{array}{c|cc}
0 & & \\
1/2 & 1/2 & \\
\hline
& 0 & 1
\end{array}
$$

The midpoint method is not the only second-order Runge–Kutta method with two stages; there is a family of such methods, parameterized by α and given by the formula

$$
y_{n+1} = y_n + h\left((1 - \tfrac{1}{2\alpha})f(t_n, y_n) + \tfrac{1}{2\alpha}f(t_n + \alpha h, y_n + \alpha hf(t_n, y_n))\right).
$$

Its Butcher tableau is

$$
\begin{array}{c|cc}
0 & & \\
\alpha & \alpha & \\
\hline
& (1 - \tfrac{1}{2\alpha}) & \tfrac{1}{2\alpha}
\end{array}
$$

In this family, $\alpha = \dfrac{1}{2}$ gives the midpoint method, and $\alpha = 1$ is Heun's method.

Usage

As an example, consider the two-stage second-order Runge–Kutta method with α = 2/3, also known as Ralston method. It is given by the tableau

$$
\begin{array}{c|cc}
0 & & \\
2/3 & 2/3 & \\
\hline
& 1/4 & 3/4
\end{array}
$$

with the corresponding equations

$$k_1 = f(t_n, y_n),$$
$$k_2 = f(t_n + \tfrac{2}{3}h,\, y_n + \tfrac{2}{3}hk_1),$$
$$y_{n+1} = y_n + h\left(\tfrac{1}{4}k_1 + \tfrac{3}{4}k_2\right).$$

This method is used to solve the initial-value problem

$$y' = \tan(y) + 1, \quad y_0 = 1, t \in [1, 1.1]$$

with step size $h = 0.025$, so the method needs to take four steps.

The method proceeds as follows:

$t_1 = 1.025$:

$$y_0 = 1$$

$$k_1 = 2.557407725$$
$$k_2 = f(t_0 + \tfrac{2}{3}h,\, y_0 + \tfrac{2}{3}hk_1) = 2.7138981184$$

$$y_1 = y_0 + h(\tfrac{1}{4}k_1 + \tfrac{3}{4}k_2) = \underline{1.066869388}$$

$t_2 = 1.05$:

$$y_1 = 1.066869388 \quad k_1 = 2.813524695 \quad k_2 = f(t_1 + \tfrac{2}{3}h,\, y_1 + \tfrac{2}{3}hk_1)$$

$$y_2 = y_1 + h(\tfrac{1}{4}k_1 + \tfrac{3}{4}k_2) = \underline{1.141332181}$$

$t_3 = 1.075$:

$$y_2 = 1.141332181 \quad k_1 = 3.183536647 \quad k_2 = f(t_2 + \tfrac{2}{3}h, \, y_2 + \tfrac{2}{3}hk_1)$$

$$y_4 = y_3 + h(\tfrac{1}{4}k_1 + \tfrac{3}{4}k_2) = \underline{1.335079087}.$$

$t_4 = 1.1:$

$$y_3 = 1.227417567 \quad k_1 = 3.796866512 \quad k_2 = f(t_3 + \tfrac{2}{3}h, \, y_3 + \tfrac{2}{3}hk_1)$$

$$y_4 = y_3 + h(\tfrac{1}{4}k_1 + \tfrac{3}{4}k_2) = \underline{1.335079087}.$$

The numerical solutions correspond to the underlined values.

Adaptive Runge–Kutta Methods

The adaptive methods are designed to produce an estimate of the local truncation error of a single Runge–Kutta step. This is done by having two methods in the tableau, one with order p nd one with order $p-1$.

The lower-order step is given by

$$y_{n+1}^{*} = y_n + h\sum_{i=1}^{s} b_i^{*} k_i,$$

where the k_i are the same as for the higher-order method. Then the error is

$$e_{n+1} = y_{n+1} - y_{n+1}^{*} = h\sum_{i=1}^{s} (b_i - b_i^{*})k_i,$$

which is $O(h^p)$. The error estimate is used to control the step size. The Butcher tableau for this kind of method is extended to give the values of b_i^{*}:

$$
\begin{array}{c|ccccc}
0 & & & & & \\
c_2 & c_2 & & & & \\
c_3 & a_{31} & a_{32} & & & \\
\vdots & \vdots & & \ddots & & \\
c_s & a_{s1} & a_{s2} & \cdots & a_{s,s-1} & \\
\hline
 & b_1 & b_2 & \cdots & b_{s-1} & b_s \\
 & b_1^{*} & b_2^{*} & \cdots & b_{s-1}^{*} & b_s^{*}
\end{array}
$$

The Runge–Kutta–Fehlberg method has two methods of orders 5 and 4. Its extended Butcher tableau is:

0						
1/4	1/4					
3/8	3/32	9/32				
12/13	1932/2197	−7200/2197	7296/2197			
1	439/216	−8	3680/513	-845/4104		
1/2	−8/27	2	−3544/2565	1859/4104	−11/40	
	16/135	0	6656/12825	28561/56430	−9/50	2/55
	25/216	0	1408/2565	2197/4104	−1/5	0

However, the simplest adaptive Runge–Kutta method involves combining Heun's method, which is order 2, with the Euler method, which is order 1. Its extended Butcher tableau is:

0		
1	1	
	1/2	1/2
	1	0

Other adaptive Runge–Kutta methods are the Bogacki–Shampine method (orders 3 and 2), the Cash–Karp method and the Dormand–Prince method (both with orders 5 and 4).

Nonconfluent Runge–Kutta Methods

A Runge–Kutta method is said to be *nonconfluent* if all the $c_i, i = 1, 2, \ldots, s$ are distinct.

Implicit Runge–Kutta Methods

All Runge–Kutta methods mentioned up to now are explicit methods. Explicit Runge–Kutta methods are generally unsuitable for the solution of stiff equations because their region of absolute stability is small; in particular, it is bounded. This issue is especially important in the solution of partial differential equations.

The instability of explicit Runge–Kutta methods motivates the development of implicit methods. An implicit Runge–Kutta method has the form

$$y_{n+1} = y_n + h\sum_{i=1}^{s} b_i k_i,$$

where

$$k_i = f\left(t_n + c_i h, \ y_n + h\sum_{j=1}^{s} a_{ij} k_j\right), \quad i = 1, \ldots, s.$$

The difference with an explicit method is that in an explicit method, the sum over j only goes up to $i - 1$. This also shows up in the Butcher tableau: the coefficient matrix a_{ij} of an explicit method is lower triangular. In an implicit method, the sum over j goes up to s and the coefficient matrix is not triangular, yielding a Butcher tableau of the form

$$
\begin{array}{c|cccc}
c_1 & a_{11} & a_{12} & \cdots & a_{1s} \\
c_2 & a_{21} & a_{22} & \cdots & a_{2s} \\
\vdots & \vdots & \vdots & \ddots & \vdots \\
c_s & a_{s1} & a_{s2} & \cdots & a_{ss} \\
\hline
 & b_1 & b_2 & \cdots & b_s \\
 & b_1^* & b_2^* & \cdots & b_s^*
\end{array}
\quad = \quad
\begin{array}{c|c}
\mathbf{c} & A \\
\hline
 & \mathbf{b}^{\mathrm{T}}
\end{array}
$$

The consequence of this difference is that at every step, a system of algebraic equations has to be solved. This increases the computational cost considerably. If a method with s stages is used to solve a differential equation with m components, then the system of algebraic equations has ms components. This can be contrasted with implicit linear multistep methods (the other big family of methods for ODEs): an implicit s-step linear multistep method needs to solve a system of algebraic equations with only m components, so the size of the system does not increase as the number of steps increases.

Examples

The simplest example of an implicit Runge–Kutta method is the backward Euler method:

$$y_{n+1} = y_n + hf(t_n + h, \ y_{n+1}).$$

The Butcher tableau for this is simply:

$$
\begin{array}{c|c}
1 & 1 \\
\hline
 & 1
\end{array}
$$

This Butcher tableau corresponds to the formulae

$$k_1 = f(t_n + h, \ y_n + hk_1) \quad and \quad y_{n+1} = y_n + hk_1,$$

which can be re-arranged to get the formula for the backward Euler method listed above.

Another example for an implicit Runge–Kutta method is the trapezoidal rule. Its Butcher tableau is:

$$
\begin{array}{c|cc}
0 & 0 & 0 \\
1 & \frac{1}{2} & \frac{1}{2} \\
\hline
 & \frac{1}{2} & \frac{1}{2} \\
 & 1 & 0
\end{array}
$$

The trapezoidal rule is a collocation method. All collocation methods are implicit Runge–Kutta methods, but not all implicit Runge–Kutta methods are collocation methods.

The Gauss–Legendre methods form a family of collocation methods based on Gauss quadrature. A Gauss–Legendre method with s stages has order $2s$ (thus, methods with arbitrarily high order can be constructed). The method with two stages (and thus order four) has Butcher tableau:

$$
\begin{array}{c|cc}
\frac{1}{2} - \frac{1}{6}\sqrt{3} & \frac{1}{4} & \frac{1}{4} - \frac{1}{6}\sqrt{3} \\
\frac{1}{2} + \frac{1}{6}\sqrt{3} & \frac{1}{4} + \frac{1}{6}\sqrt{3} & \frac{1}{4} \\
\hline
 & \frac{1}{2} & \frac{1}{2} \\
 & \frac{1}{2} + \frac{1}{2}\sqrt{3} & \frac{1}{2} - \frac{1}{2}\sqrt{3}
\end{array}
$$

Stability

The advantage of implicit Runge–Kutta methods over explicit ones is their greater stability, especially when applied to stiff equations. Consider the linear test equation $y' = \lambda y$. A Runge–Kutta method applied to this equation reduces to the iteration $y_{n+1} = r(h\lambda)y_n$, with r given by

$$
r(z) = 1 + zb^T (I - zA)^{-1} e = \frac{\det(I - zA + zeb^T)}{\det(I - zA)},
$$

where e stands for the vector of ones. The function r is called the *stability function*. It follows from the formula that r is the quotient of two polynomials of degree s if the method has s stages. Explicit methods have a strictly lower triangular matrix A, which implies that $\det(I - zA) = 1$ and that the stability function is a polynomial.

The numerical solution to the linear test equation decays to zero if $|r(z)| < 1$ with $z = h\lambda$. The set of such z is called the *domain of absolute stability*. In particular, the method is said to be A-stable if all z with $\mathrm{Re}(z) < 0$ are in the domain of absolute stability. The stability function of an explicit Runge–Kutta method is a polynomial, so explicit Runge–Kutta methods can never be A-stable.

If the method has order p, then the stability function satisfies $r(z) = e^z + O(z^{p+1})$ as

$z \to 0$. Thus, it is of interest to study quotients of polynomials of given degrees that approximate the exponential function the best. These are known as Padé approximants. A Padé approximant with numerator of degree m and denominator of degree n is A-stable if and only if $m \le n \le m + 2$.

The Gauss–Legendre method with s stages has order $2s$, so its stability function is the Padé approximant with $m = n = s$. It follows that the method is A-stable. This shows that A-stable Runge–Kutta can have arbitrarily high order. In contrast, the order of A-stable linear multistep methods cannot exceed two.

B-stability

The *A-stability* concept for the solution of differential equations is related to the linear autonomous equation $y' = \lambda y$. Dahlquist proposed the investigation of stability of numerical schemes when applied to nonlinear systems that satisfy a monotonicity condition. The corresponding concepts were defined as *G-stability* for multistep methods (and the related one-leg methods) and *B-stability* (Butcher, 1975) for Runge–Kutta methods. A Runge–Kutta method applied to the non-linear system $y' = f(y)$, which verifies $\langle f(y) - f(z), y - z \rangle < 0$, is called *B-stable*, if this condition implies $\| y_{n+1} - z_{n+1} \| \le \| y_n - z_n \|$ for two numerical solutions.

Let B, M and Q be three $s \times s$ matrices defined by

$$B = diag(b_1, b_2, \ldots, b_s), M = BA + A^T B - bb^T, Q = BA^{-1} + A^{-T} B - A^{-T} bb^T A^{-1}.$$

A Runge–Kutta method is said to be *algebraically stable* if the matrices B and M are both non-negative definite. A sufficient condition for *B-stability* is: B and Q are non-negative definite.

Derivation of the Runge–Kutta fourth-order Method

In general a Runge–Kutta method of order s can be written as:

$$y_{t+h} = y_t + h \cdot \sum_{i=1}^{s} a_i k_i + \mathcal{O}(h^{s+1}),$$

where:

$$k_i = f\left(y_t + h \cdot \sum_{j=1}^{s} \beta_{ij} k_j, t_n + \alpha_i h \right)$$

are increments obtained evaluating the derivatives of   at the i-th order.

We develop the derivation for the Runge–Kutta fourth-order method using the general formula with $s = 4$ evaluated, as explained above, at the starting point, the midpoint and the end point of any interval $(t, t + h)$, thus we choose:

$$\alpha_i \qquad\qquad \beta_{ij}$$

$$\alpha_1 = 0 \qquad \beta_{21} = \frac{1}{2}$$

$$\alpha_2 = \frac{1}{2} \qquad \beta_{32} = \frac{1}{2}$$

$$\alpha_3 = \frac{1}{2} \qquad \beta_{43} = 1$$

$$\alpha_4 = 1$$
$$\beta_{43} = 1$$

and $\beta_{ij} = 0$ otherwise. We begin by defining the following quantities:

$$y_{t+h}^1 = y_t + hf(y_t, t)$$

$$y_{t+h}^2 = y_t + hf\left(y_{t+h/2}^1, t + \frac{h}{2}\right)$$

$$y_{t+h}^3 = y_t + hf\left(y_{t+h/2}^2, t + \frac{h}{2}\right)$$

where $y_{t+h/2}^1 = \dfrac{y_t + y_{t+h}^1}{2}$ and $y_{t+h/2}^2 = \dfrac{y_t + y_{t+h}^2}{2}$ If we define:

$$k_1 = f(y_t, t)$$

$$k_2 = f\left(y_{t+h/2}^1, t + \frac{h}{2}\right)$$

$$k_3 = f\left(y_{t+h/2}^2, t + \frac{h}{2}\right)$$

$$k_4 = f\left(y_{t+h}^3, t + h\right)$$

and for the previous relations we can show that the following equalities holds up to $\mathcal{O}(h^2)$:

$$k_2 = f\left(y_{t+h/2}^1, t+\frac{h}{2}\right) = f\left(y_t + \frac{h}{2}k_1, t+\frac{h}{2}\right) = f(y_t, t) + \frac{h}{2}\frac{d}{dt}f(y_t, t)$$

$$k_3 = f\left(y_{t+h/2}^2, t+\frac{h}{2}\right) = f\left(y_t + \frac{h}{2}f\left(y_t + \frac{h}{2}k_1, t+\frac{h}{2}\right), t+\frac{h}{2}\right) = f(y_t, t) + \frac{h}{2}\frac{d}{dt}\left[f(y_t, t) + \frac{h}{2}\frac{d}{dt}f(y_t, t)\right]$$

$$k_4 = f\left(y_{t+h}^3, t+h\right) = f\left(y_t + hf\left(y_t + \frac{h}{2}k_2, t+\frac{h}{2}\right), t+h\right) = f\left(y_t + hf\left(y_t + \frac{h}{2}f\left(y_t + \frac{h}{2}f(y_t, t), t+\frac{h}{2}\right), t+\frac{h}{2}\right), t+h\right)$$

$$= f(y_t, t) + h\frac{d}{dt}\left[f(y_t, t) + \frac{h}{2}\frac{d}{dt}\left[f(y_t, t) + \frac{h}{2}\frac{d}{dt}f(y_t, t)\right]\right]$$

where:

$$\frac{d}{dt}f(y_t,t) = \frac{\partial}{\partial y}f(y_t,t)\dot{y}_t + \frac{\partial}{\partial t}f(y_t,t) = f_y(y_t,t)\dot{y} + f_t(y_t,t) := \ddot{y}_t$$

is the total derivative of f with respect to time.

If we now express the general formula using what we just derived we obtain:

$$y_{t+h} = y_t + h\left\{ a \cdot f(y_t,\ t) + b \cdot \left[f(y_t,\ t) + \frac{h}{2}\frac{d}{dt}f(y_t,\ t) \right] + \right.$$

$$+ c \cdot \left[f(y_t,\ t) + \frac{h}{2}\frac{d}{dt}\left[f(y_t,\ t) + \frac{h}{2}\frac{d}{dt}f(y_t,\ t) \right] \right] +$$

$$+ d \cdot \left[f(y_t,\ t) + h\frac{d}{dt}\left[f(y_t,\ t) + \frac{h}{2}\frac{d}{dt}\left[f(y_t,\ t) + \frac{h}{2}\frac{d}{dt}f(y_t,\ t) \right] \right] \right.$$

$$= y_t + a \cdot hf_t + b \cdot hf_t + b \cdot \frac{h^2}{2}\frac{df_t}{dt} + c \cdot hf_t + c \cdot \frac{h^2}{2}\frac{df_t}{dt} +$$

$$+ c \cdot \frac{h^3}{4}\frac{d^2 f_t}{dt^2} + d \cdot hf_t + d \cdot h^2\frac{df_t}{dt} + d \cdot \frac{h^3}{2}\frac{d^2 f_t}{dt^2} + d \cdot \frac{h^4}{4}\frac{d^3 f_t}{dt^3} +$$

and comparing this with the Taylor series of y_{t+h} around y_t:

$$y_{t+h} = y_t + h\dot{y}_t + \frac{h^2}{2}\ddot{y}_t + \frac{h^3}{6}y_t^{(3)} + \frac{h^4}{24}y_t^{(4)} + \mathcal{O}(h^5) =$$

$$= y_t + hf(y_t,t) + \frac{h^2}{2}\frac{d}{dt}f(y_t,t) + \frac{h^3}{6}\frac{d^2}{dt^2}f(y_t,t) + \frac{h^4}{24}\frac{d^3}{dt^3}f(y_t,t)$$

we obtain a system of constraints on the coefficients:

$$\begin{cases} a+b+c+d = 1 \\[1mm] \dfrac{1}{2}b + \dfrac{1}{2}c + d = \dfrac{1}{2} \\[1mm] \dfrac{1}{4}c + \dfrac{1}{2}d = \dfrac{1}{6} \\[1mm] \dfrac{1}{4}d = \dfrac{1}{24} \end{cases}$$

which when solved gives $a = \dfrac{1}{6}, b = \dfrac{1}{3}, c = \dfrac{1}{3}, d = \dfrac{1}{6}$ as stated above.

System with Two DOF

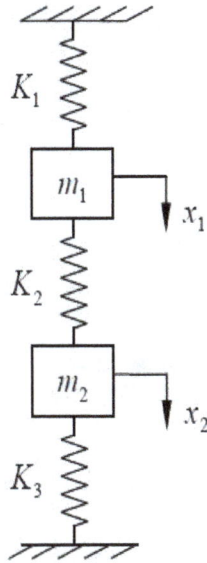

Let us assume that the two degrees of freedoms are measured as x_1 and x_2 as indicated in the figure.

A two DOF system has 2 natural frequencies. When a system vibrates at any natural frequencythere exists a deflection relationship with respect to amplitudes.

Such a configuration is called normal mode. In a normal mode oscillation both m_1 and m_2 have harmonic motion of same frequency passing through simultaneously through the equilibrium position.

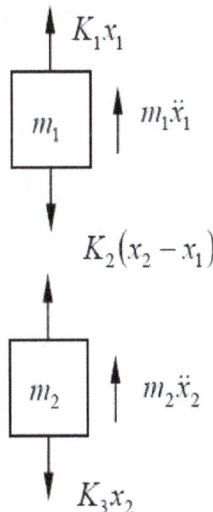

Considering the equilibrium of

Mass $m_1 \Rightarrow m_1 \ddot{x}_1 + K_1 x_1 - K(x_2 - x_1) = 0$

or, $m_1 \ddot{x}_1 + (K_1 + K_2) x_1 - K_2 x_2 = 0 \rightarrow (1)$

Equilibrium of mass $m_2 \Rightarrow m_2 \ddot{x}_2 + K_2(x_2 - x_1) + K_3 x_2 = 0$

or, $m_2 \ddot{x}_2 + K_2 x_1 + (K_2 + K_3) x_2 = 0 \rightarrow (2)$

Express (1) & (2) in matrix form

$$\begin{bmatrix} m_1 & 0 \\ 0 & m_2 \end{bmatrix} \begin{Bmatrix} \ddot{x}_1 \\ \ddot{x}_2 \end{Bmatrix} + \begin{bmatrix} K_1 + K_2 & -K_2 \\ -K_2 & K_2 + K_3 \end{bmatrix} \begin{Bmatrix} x_1 \\ x_2 \end{Bmatrix} = 0$$

Assuming m_1 & m_2 oscillating at the same frequancy, but at different amplitudes.

$$x_1 = A_1 \sin(\omega t + \alpha)$$

$$x_2 = A_2 \sin(\omega t + \alpha) \rightarrow (3)$$

$$\left. \begin{aligned} \Rightarrow x_1 &= A e^{i\omega t} \\ x_2 &= A_2 e^{i\omega t} \end{aligned} \right\} alternative\ form$$

Substituting (3) in matrix equation

$$-m_1 A_1 \omega^2 + (K_1 + K_2) A_1 - K_2 A_2 = 0$$

$$-m_2 A_2 \omega^2 + (K_2 + K_3) A_2 - K_2 A_1 = 0$$

$$\Rightarrow \begin{bmatrix} -\omega^2 + \dfrac{K_1 + K_2}{m_1} & -\dfrac{K_2}{m_1} \\ -\dfrac{K_2}{m_2} & -\omega^2 + \dfrac{K_2 + K_3}{m_2} \end{bmatrix} \begin{Bmatrix} A_1 \\ A_2 \end{Bmatrix} = 0$$

For non-trivial solution

$$\begin{vmatrix} -\omega^2 + \dfrac{K_1 + K_2}{m_1} & -\dfrac{K_2}{m_1} \\ -\dfrac{K_2}{m_2} & -\omega^2 + \dfrac{K_2 + K_3}{m_2} \end{vmatrix} = 0$$

$$\Rightarrow \quad \omega^4 - \omega^2 \left[\frac{K_1+K_2}{m_1} + \frac{K_2+K_3}{m_2} \right] + \left[\left(\frac{K_1+K_2}{m_1} \right)\left(\frac{K_2+K_3}{m_2} \right) - \frac{K_2^2}{m_1 m_2} \right] = 0$$

$$\omega_{1,2}^2 = \frac{1}{2}\left[\frac{K_1+K_2}{m_1} + \frac{K_2+K_3}{m_2} \right]$$

$$\pm \frac{1}{2}\left[\left\{ \frac{K_1+K_2}{m_1} + \frac{K_2+K_3}{m_2} \right\}^2 - 4\left\{ \left(\frac{K_1+K_2}{m_1} \right)\left(\frac{K_2+K_3}{m_2} \right) - \right\} \frac{K_2^2}{m_1 m_2} \right]^{\frac{1}{2}}$$

$$= \frac{1}{2}\left[\frac{K_1+K_2}{m_1} + \frac{K_2+K_3}{m_2} \right] \pm \frac{1}{2}\left[\left\{ \frac{K_1+K_2}{m_1} - \frac{K_2+K_3}{m_2} \right\}^2 + \frac{4K_2^2}{m_1 m_2} \right]^{\frac{1}{2}} \quad \therefore (a-b)^2 = (a+b)^2 - 4ab$$

$$\left(\frac{A_1}{A_2} \right)_{\omega_1^2} = \frac{-K_2}{m_1 \omega_1^2 - K_1 - K_2} = \frac{m_2 \omega_2^2 - K_2 - K_3}{-K_2}$$

$$\left(\frac{A_1}{A_2} \right)_{\omega_2^2} = \frac{-K_2}{m_1 \omega_2^2 - K_1 - K_2} = \frac{m_2 \omega_2^2 - K_2 - K_3}{-K_2}$$

Problem 1 Find out the mode shapes of the following spring mass systems.

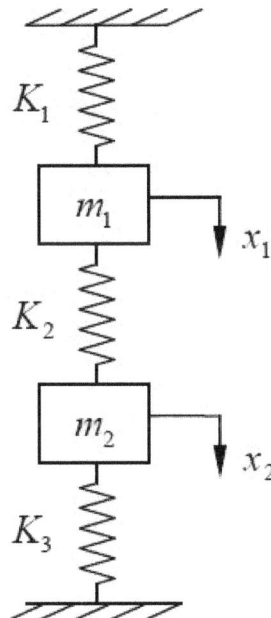

$$\omega_1^2 = \frac{K}{m} \qquad \omega_2^2 = \frac{K+2K_2}{m} \qquad \begin{aligned} \omega_1 &\rightarrow \begin{Bmatrix} 1 \\ 1 \end{Bmatrix} \\ \omega_2 &\rightarrow \begin{Bmatrix} 1 \\ -1 \end{Bmatrix} \end{aligned}$$

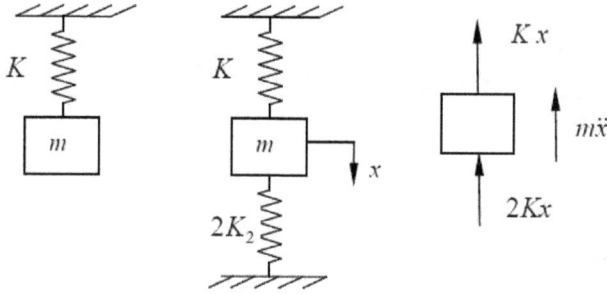

$$\left|\frac{A_1}{A_2}\right|_{\omega_1^2 = \frac{K}{m}} = 1 \quad ; \quad \left|\frac{A_1}{A_2}\right|_{\omega_2^2 = \frac{K+2K_2}{m}} = -1$$

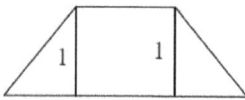

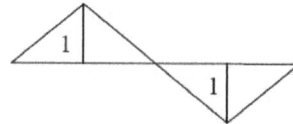

Two masses move in the same
direction through the same distance

Two masses move through same
distance but in the opposite direction.

Problem 2

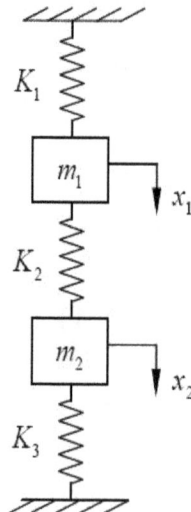

$$K_1 = K_2 - K_3 = K$$

$$m_1 = m \quad ; \quad m_2 = 2m$$

$$\omega^4 - \omega^2 \left(\frac{3K}{m} \right) + \frac{3}{2} \left(\frac{K}{m} \right)^2 = 0$$

$$\Rightarrow \left. \begin{array}{l} \omega_1^2 = 0.634 \dfrac{K}{m} \\[2mm] \omega_2^2 = 2.366 \dfrac{K}{m} \end{array} \right\}$$

$$\left(\frac{A_1}{A_2} \right)_{\omega_1^2} = \frac{K}{2K - m\omega_1^2} = 0.731 \qquad \Rightarrow$$

$$\left(\frac{A_1}{A_2} \right)_{\omega_2^2} = \frac{K}{2K - m\omega_2^2} = -2.73 \qquad \Rightarrow$$

Problem 3

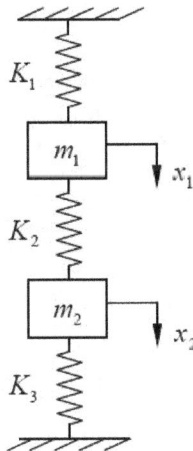

$$K_1 = K_2 = K \qquad K_3 = 0$$

$$m_1 = m_2 = m_3 \qquad \rightarrow \qquad \text{Draw mode shape}$$

Problem 4

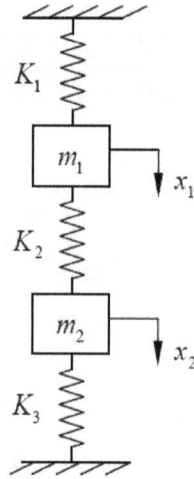

$$\left.\begin{array}{l} K_1 = K_3 = K \\ K_2 = nK \end{array}\right\} \qquad \begin{array}{l} m_1 = m \\ m_2 = 2m \end{array}$$

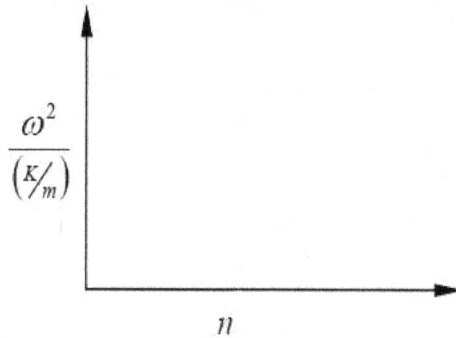

$$\text{Plot } (n = 2, \ n = 4) \ \rightarrow \ \text{mode shape}$$

General solution

First Mode $\rightarrow$ oscillating at ω_1

$$x_1 = A_1 \sin(\omega_1 t + \alpha_1)$$
$$x_2 = \beta_1 A_1 \sin(\omega_1 t + \alpha_1)$$

$$\text{where } \beta_1 = \left(\frac{A_2}{A_1}\right)_{\omega_1^2}$$

Second Mode $\rightarrow$ oscillating at ω_2

$$x_1 = A_2 \sin(\omega_2 t + \alpha_2)$$
$$x_2 = \beta_2 A_2 \sin(\omega_2 t + \alpha_2)$$

$$\text{where } \beta_2 = \left(\frac{A_2}{A_1}\right)_{\omega_2^2}$$

General Solution

$$x_1 = A_1 \sin(\omega_1 t + \alpha_1) + A_2 \sin(\omega_2 t + \alpha_2)$$

$$x_1 = \beta_1 A_1 \sin(\omega_1 t + \alpha_1) + \beta_2 A_2 \sin(\omega_2 t + \alpha_2)$$

$\omega_1 \,\&\, \omega_2 \rightarrow$ frequencies as given earlier

$$\beta_1 = \left(\frac{A_2}{A_1}\right)_{\omega_1^2} = \frac{m_1 \omega_1^2 - K - K_2}{-K_2}$$

$$\beta_2 = \left(\frac{A_2}{A_1}\right)_{\omega_2^2} = \frac{m_1 \omega_2^2 - K_1 - K_2}{-K_2}$$

Where $A_1, A_2, \alpha_1, \alpha_2$ are determined from initial conditions of displacements and velocities.

$$x_1, x_2, \dot{x}_1, \dot{x}_2 \text{ at } t = 0$$

Problem 5

$$\begin{array}{l} K_1 = K_2 = K_3 = K \\ m_1 = m_2 = m \end{array} \Bigg\} \quad \begin{array}{l} x_1(0) = 5 \\ x_2(0) = 0 \end{array} \qquad \dot{x}(0) = \dot{x}(0) = 0$$

$$\omega_1^2 = \frac{K}{m} \qquad \omega_2^2 = \frac{3K}{m}$$

$$\left(\frac{A_1}{A_2}\right)_{\omega_1^2} = 1 \quad ; \quad \beta_1 = 1$$

$$\left(\frac{A_1}{A_2}\right)_{\omega_2^2} = -1 \quad ; \quad \beta_2 = -1$$

Initial conditions (translation)

$$\left. \begin{array}{l} 5 = A_1 \sin \alpha_1 + A_2 \sin \alpha_2 \\ 0 = A_1 \sin \alpha_1 - A_2 \sin \alpha_2 \end{array} \right\} \qquad \begin{array}{l} A_1 \sin \alpha_1 = 2.5 \\ A_2 \sin \alpha_2 = 2.5 \end{array}$$

Initial conditian (velocity)

$$0 = \omega_1 A_1 \cos\alpha_1 + \omega_2 A_2 \cos a_2 \quad \rightarrow \cos a_1 = 0 \Rightarrow a_1 = 90^0$$

$$0 = \omega_1 A_1 \cos\alpha_1 - \omega_2 A_2 \cos a_2 \quad \rightarrow \cos a_2 = 0 \Rightarrow a_2 = 90^0$$

$$\therefore A_1 = 2.5; A_2 = 2.5$$

Solution

$$x_1 = 2.5\cos\sqrt{\frac{K}{m}}t + 2.5\cos\sqrt{\frac{3K}{m}}t$$

$$x_2 = 2.5\cos\sqrt{\frac{K}{m}}t - 2.5\cos\sqrt{\frac{3K}{m}}t$$

Beat Phenomenon

A very interesting phenomenon is encountered when the natural frequencies of a two degree – of – freedom system are very close in value. To illustrate the phenomenon, let us consider two identical pendulums connected by a spring. The corresponding free – body are shown in which the assumption of small angles θ_1 and θ_2 is implied. The moment equations about the points O and O', respectively, yield the differential equations of motion

$$mL^2\theta_1' + mgL\theta_1 + ka^2\left(\theta_1 - \theta_2\right) = 0$$

$$mL^2\theta_2' + mgL\theta_2 - ka^2\left(\theta_1 - \theta_2\right) = 0$$

Which can be arranged in the matrix form

$$\begin{bmatrix} mL^2 & 0 \\ 0 & mL^2 \end{bmatrix}\begin{Bmatrix} \theta_1' \\ \theta_2' \end{Bmatrix} + \begin{bmatrix} mhL + ka^2 & -ka^2 \\ -ka^2 & mgL + ka^2 \end{bmatrix}\begin{Bmatrix} \theta_1 \\ \theta_2 \end{Bmatrix} = \begin{Bmatrix} 0 \\ 0 \end{Bmatrix}$$

Indicating that the system is coupled elastically. As expected, when the spring stiffness k reduces to zero the coupling disappears and the two pendulums reduce to independent simple pendulums with identical natural frequencies equal to $\sqrt{g/L}$. For $k \neq 0$, Equation above yields the Eigen value problem.

$$-\omega^2\begin{bmatrix} mL^2 & 0 \\ 0 & mL^2 \end{bmatrix}\begin{Bmatrix} \Theta_1 \\ \Theta_2 \end{Bmatrix} + \begin{bmatrix} mgL + ka^2 & -ka^2 \\ -ka^2 & mgL + ka^2 \end{bmatrix}\begin{Bmatrix} \Theta_1 \\ \Theta_2 \end{Bmatrix} = \begin{Bmatrix} 0 \\ 0 \end{Bmatrix}$$

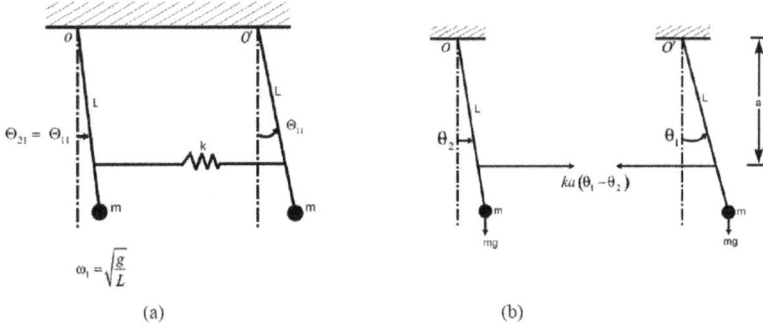

(a)   (b)

leading to the characteristic equation

$$\det \begin{bmatrix} mgL + ka^2 - \omega^2 mL^2 & -ka^2 \\ -ka^2 & mgL + ka^2 - \omega^2 mL^2 \end{bmatrix}$$

$$= \left(mgL + ka^2 - \omega^2 mL^2 \right)^2 - \left(ka^2 \right)^2 = 0$$

Which is equivalent to

$$mgL + ka^2 - \omega^2 mL^2 = \pm ka^2$$

Hence the two natural frequencies are

$$\omega_1 = \sqrt{\frac{g}{L}} \omega_2 = \sqrt{\frac{g}{L} + 2\frac{k}{m}\frac{a^2}{L^2}}$$

The natural modes are obtained from the equations

$$-\omega_i^2 \begin{bmatrix} mL^2 & 0 \\ 0 & mL^2 \end{bmatrix} \begin{Bmatrix} \Theta_1 \\ \Theta_2 \end{Bmatrix}_i = \begin{Bmatrix} 0 \\ 0 \end{Bmatrix} \quad i = 1, 2$$

Inserting $\omega_1^2 = g/L$ and $\omega_2^2 = g/L + 2(k/m)(a^2/L^2)$ into equation above , and solving for the Ratios Θ_{21}/Θ_{11} and Θ_{22}/Θ_{12}, we obtain

$$\frac{\Theta_{21}}{\Theta_{11}} = 1 \qquad \frac{\Theta_{22}}{\Theta_{12}} = -1$$

So that in the first natural mode the two pendulums move like a single pendulum with the spring k unstretched, which can also be concluded from the fact that the first natural frequency of the system is that of the simple pendulum, $\omega_1 = \sqrt{g/L}$. On the other hand, in the second natural mode the two pendulums are 180° out of phase.

As was pointed out, the general motion of the system can be expressed as a superposition of the two natural modes multiplied by the associated

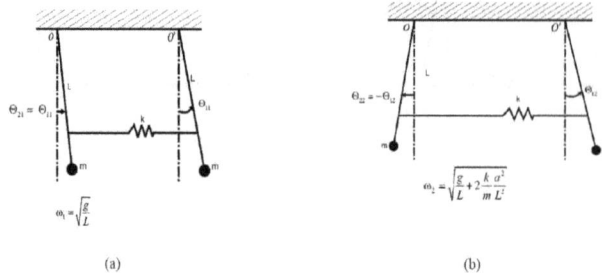

(a)   (b)

Natural coordinates, or

$$\begin{Bmatrix} \theta_1(t) \\ \theta_2(t) \end{Bmatrix} = C_1 \begin{Bmatrix} \Theta_1 \\ \Theta_2 \end{Bmatrix}_1 \cos(\omega_1 t - \phi_1) + C_2 \begin{Bmatrix} \Theta_1 \\ \Theta_2 \end{Bmatrix}_2 \cos(\omega_2 t - \phi_2)$$

Choosing $\Theta_{11} = \Theta_{12} = 1$ and using the equations above it can be rewritten in the scalar form.

$$\theta_1(t) = C_1 \cos(\omega_1 t - \phi_1) + C_2 \cos(\omega_2 t - \phi_2)$$

$$\theta_2(t) = C_1 \cos(\omega_1 t - \phi_1) + C_2 \cos(\omega_2 t - \phi_2)$$

Letting the initial conditions be $\theta_1(0) = \theta_0$, $\theta_2(0) = \dot{\theta}_1(0) = \dot{\theta}_2(0) = 0$ Equation above becomes

$$\theta_1(t) = \frac{1}{2}\theta_0 \cos \omega_1 t + \frac{1}{2}\theta_0 \cos \omega_2 t = \theta_0 \cos \frac{\omega_2 - \omega_1}{2} t \cos \frac{\omega_2 + \omega_1}{2} t$$

$$\theta_2(t) = \frac{1}{2}\theta_0 \cos \omega_1 t - \frac{1}{2}\theta_0 \cos \omega_2 t$$

$$= \theta_0 \sin \frac{\omega_2 - \omega_1}{2} t \sin \frac{\omega_2 + \omega_1}{2} t$$

Note that, in deriving the above equation, we used the trigonometric relations,

$$\cos(\alpha \pm \beta) = \cos \alpha \cos \beta \pm \sin \alpha \sin \beta \text{ in which } \alpha = (\omega_2 - \omega_1)t/2, \beta = (\omega_2 + \omega_1)t/2$$

Next let us consider the case in which ka² is very small in value compared withmqL . Examining the second equation under the given topic, we conclude that this statement is equivalent to saying that the coupling provided by the spring k is very weak. In this case, the above equation can be written in the form:

$$\theta_1(t) \cong \theta_0 \cos\frac{1}{2}\omega_B t \cos\omega_{ave}t$$

$$\theta_2(t) \cong \theta_0 \sin\frac{1}{2}\omega_B t \sin\omega_{ave}t$$

Where $\omega_B/2$ and ω_{ave} are approximated by

$$\frac{\omega_B}{2} = \frac{\omega_2 - \omega_1}{2} \cong \frac{1}{2}\frac{k}{m}\frac{a^2}{\sqrt{gL^3}} \qquad \omega_{ave} = \frac{\omega_2 - \omega_1}{2} \cong \sqrt{\frac{g}{L} + \frac{1}{2}\frac{k}{m}\frac{a^2}{\sqrt{gL^3}}}$$

Hence, $\theta_1(t)$ and $\theta_2(t)$ can be regarded as being harmonic functions with frequency ω_{ave} and with amplitudes varying slowly according to $\theta_0 \cos\frac{1}{2}\omega_B t$ and $\theta_0 \sin\frac{1}{2}\omega_B t$ respectively. The plots $\theta_1(t)$ versus t and $\theta_2(t)$ versus t are shown in the figure below with the slowly varying amplitudes indicated by the dashed – line envelopes.

implies that the two harmonic functions possessing equal amplitudes and nearly equal frequencies are added, then the resulting function is an amplitude – modulated harmonic function with a frequency equal to the average frequency. At first, when the two harmonic waves reinforce each other, the amplitude is doubled, and later as the two waves cancels each other.

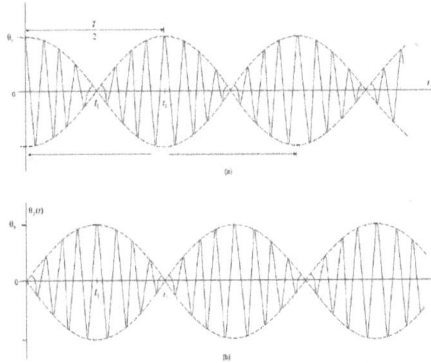

The amplitude reduces to zero. The phenomenon is known as the beat phenomenon, and the requency of modulation ω_B, which in this particular case is equal to $ka^2/m\sqrt{gL^3}$ is called the beat frequency. From the above figure a we conclude that the time between two maxima is $T/2 = 2\pi/\omega_B$, whereas the period of the amplitude – modulated envelope is $T = 4\pi/\omega_B$.

Although in our particular case the beat phenomenon resulted from the weak coupling of two pendulums, the phenomenon is not exclusively associated with two – degree –

of – freedom systems. Indeed, the beat phenomenon is purely the result of adding two harmonic functions of equal amplitudes and nearly equal frequencies. For example, the phenomenon occurs in twin – engine propeller airplanes, in which the propeller noise grows and diminishes in intensity as the sound waves generated by the two propellers reinforce and cancel each other in turn.

We observe from the above figure that there is a $90°$ phase angle between $\theta_1(t)$ and $\theta_2(t)$. At $t = 0$ the first pendulum (right pendulum in Fig. a) begins to swing with the amplitude θ_0 while the second pendulum is at rest. Soon thereafter the second pendulum is entrained, gaining amplitude while the amplitude of the first decreases. At $t_1 = \pi / \omega_B$ the amplitude of the first pendulum becomes zero, whereas the amplitude of the second pendulum reaches θ_0. At $t_2 = 2\pi / \omega_b$ the amplitude of the

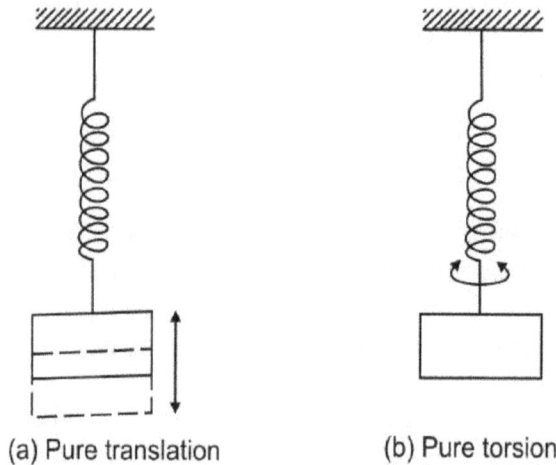

(a) Pure translation   (b) Pure torsion

First pendulum reaches θ_0 once again and that of the second pendulum reduces to zero. The motion keeps repeating itself, so that every interval of time $T / 4 = \pi / \omega_B$ there is a complete transfer of energy from one pendulum to the other.

Another example of a system exhibiting the beat phenomenon is the "Wilberforce Spring", consisting of a mass of finite dimensions suspended by a helical spring such that the frequency of vertical translation and the frequency of torsional motion are very close in value. In this case, the kinetic energy changes from pure translational in the vertical direction to pure rotational about the vertical axis, as shown in the figure above.

Co-ordinate Coupling:

$$\begin{bmatrix} m_{11} & m_{12} \\ m_{21} & m_{22} \end{bmatrix} \begin{Bmatrix} \ddot{x}_1 \\ \ddot{x}_2 \end{Bmatrix} + \begin{bmatrix} K_{11} & K_{12} \\ K_{21} & K_{22} \end{bmatrix} \begin{Bmatrix} x_1 \\ x_2 \end{Bmatrix} = \begin{Bmatrix} 0 \\ 0 \end{Bmatrix}$$ - most general case of coupling

Static Coupling $(m_{12} = m_{21} = 0)$ when mass matrix is diagonal

Mass /Dynamics Coupling:

When, mass matrix is non-diagonal.

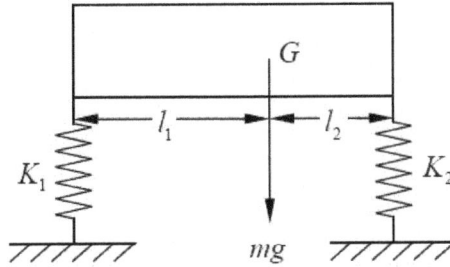

G – Centre of mass

K_1, K_2 – Spring Constants

Co-ordinatex & θ

After given a displacement and rotation,

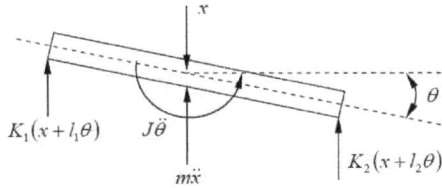

Assume θ is small $\rightarrow \therefore \cos\theta = 1$

J = mass moment of inertia about a lateral axis passing through G.

$$= mr^2$$

$$\sum F_x = 0 \qquad \sum M_G = 0$$

$$\Rightarrow m\ddot{x} + K_1\left(x - l_1\theta\right) + K_2\left(x + l_2\theta\right) = 0$$

$$\Rightarrow J\ddot{\theta} + K_2\left(x + l_2\theta\right)l_2\cos\theta - K_1\left(x - l_1\theta\right)l_1\cos\theta = 0$$

since, $\cos\theta = 1$ (for a small value of θ)

$$\Rightarrow \begin{bmatrix} m & 0 \\ 0 & J \end{bmatrix}\begin{Bmatrix} \ddot{x} \\ \ddot{\theta} \end{Bmatrix} + \begin{bmatrix} K_1 + K_2 & K_2l_2 - K_1l_1 \\ K_2l_2 - K_1l & K_1l_1^2 + K_2l_2^2 \end{bmatrix}\begin{Bmatrix} x \\ 0 \end{Bmatrix} = 0$$

static coupling case

If $K_1 l_1 = K_2 l_2 \rightarrow$ no coupling (2 independant co-ordinate)

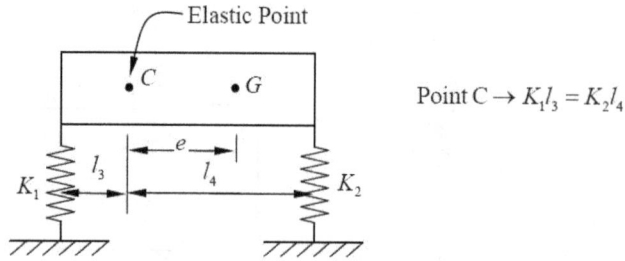

Point C $\rightarrow K_1 l_3 = K_2 l_4$

When the force applied normal to a bar produces translation. Assume x and θ are independent co-ordinate at elastic point C.

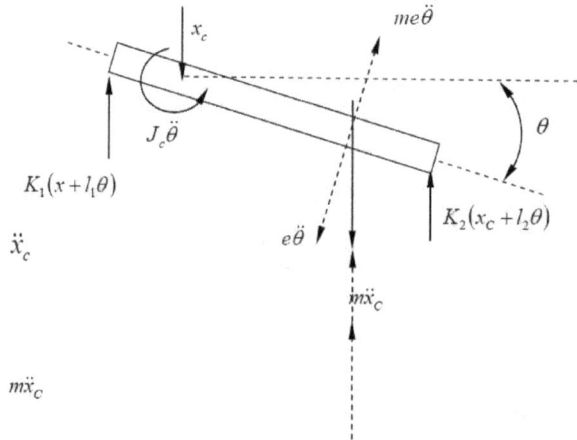

[Mass travels a distance of $e \tan \theta \approx e\theta$]

$$\sum F_x = 0 \Rightarrow m\ddot{x}_c + me\ddot{\theta}\cos\theta + K_1(x_c - l_3\theta) + K_2(x_c + l_4\theta) = 0$$

$$\sum M_c = 0 \Rightarrow (m\ddot{x}_C)e + me^2\ddot{\theta}\cos\theta + J_C\ddot{\theta} - K_1(x_c - l_3\theta)l_3 + K_2(x_C + l_4\theta)l_4\cos\theta = 0$$

Since, $\cos\theta \approx 1$

$$\Rightarrow \begin{bmatrix} m & me \\ me & J_c + me^2 \end{bmatrix}\begin{Bmatrix} \ddot{x}_C \\ \ddot{\theta} \end{Bmatrix} + \begin{bmatrix} K_1 + K_2 & K_2 l_4 - K_1 l_3 \\ K_2 l_4 - K_1 l_3 & K_1 l_3^2 + K_2 l_4^2 \end{bmatrix}\begin{Bmatrix} x_C \\ \theta \end{Bmatrix} = \{0\}$$

Where, $K_2 l_4 = K_1 l_3 \leftarrow$ dynamic coupling exit.

Say we choose the origin at the left hand of the ber.

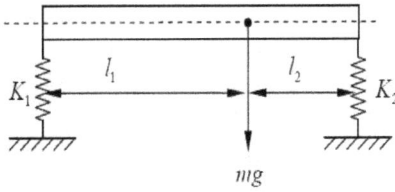

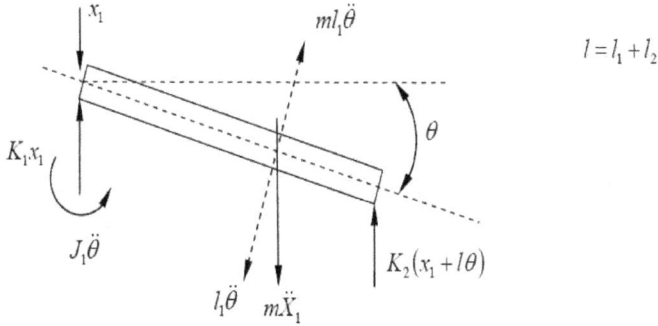

$$l = l_1 + l_2$$

Co-ordinate

$$\sum F_X = 0 \Rightarrow m\ddot{X}_1 + ml_1\ddot{\theta}\cos\theta + K_1 x_1 + K_2(x_1 + l\theta) = 0$$

$$\sum M_1 = 0 \Rightarrow m\ddot{X}_1 l_1 \cos\theta + J_1\ddot{\theta} + K_2(x_1 + l\theta)l\cos\theta + ml_1^2\ddot{\theta}\cos\theta = 0$$

$$\Rightarrow \begin{bmatrix} m & ml_1 \\ ml_1 & J_1 + ml_1^2 \end{bmatrix} \begin{Bmatrix} \ddot{x} \\ \ddot{\theta} \end{Bmatrix} + \begin{bmatrix} K_1 + K_2 & K_2 l \\ K_2 l & K_1 l^2 \end{bmatrix} \begin{Bmatrix} x \\ \theta \end{Bmatrix} = \{0\}$$

Problem:

(we may have find some other set of mechanical values)

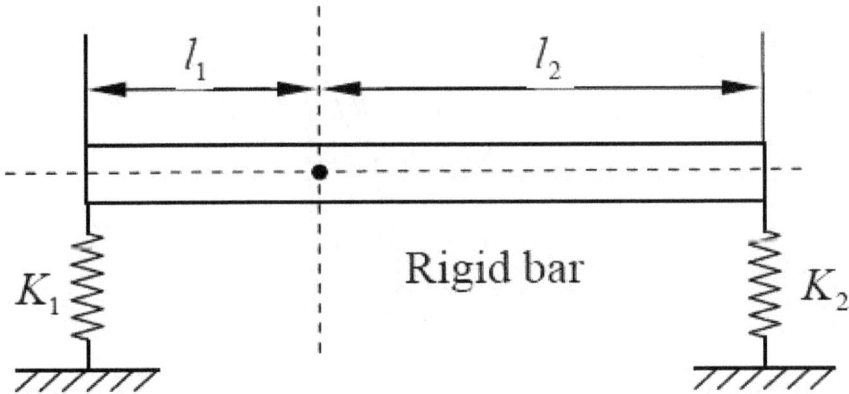

$$l = l_1 + l_2$$
$$w = 14323N \qquad l_1 = 1.3716m$$
$$g = 9.81m/s^2 \quad l_2 = 1.6764m$$
$$l = 3.04m \qquad r = 37944N/m$$
$$J_G = \frac{w}{g}(r^2) = 2170kg - m$$
$$K_1 = 35025\,M/m$$
$$K_2 = 37944N/m$$

Determine the normal modes of vibration → Frequencies and mode shapes

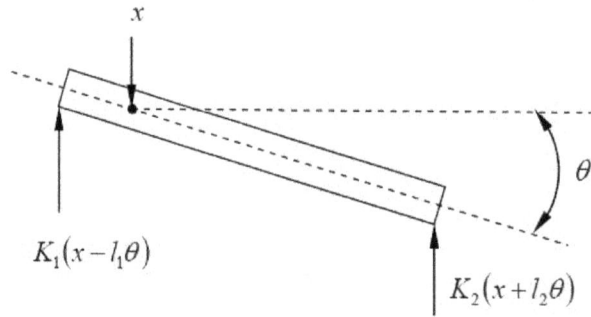

$$K_1(x - l_1\theta)$$
$$K_2(x + l_2\theta)$$

$$\Rightarrow \begin{bmatrix} m & 0 \\ 0 & J_g \end{bmatrix} \begin{Bmatrix} \ddot{x} \\ \ddot{\theta} \end{Bmatrix} + \begin{bmatrix} K_1 + K_2 & K_2 l_2 - K_1 l_1 \\ K_2 l_2 - K_1 l_1 & K_1 l_1^2 + K_2 l_2^2 \end{bmatrix} \begin{Bmatrix} x \\ \theta \end{Bmatrix} = 0 \rightarrow$$

For harmonic oscilations

$$x = Ae^{i\omega t} \qquad\qquad \theta = \theta_0 e^{i\omega t} \quad \rightarrow$$

Substituting above equations

$$\begin{bmatrix} K_1 + K_2 - \omega^2 m & K_2 l_2 - K_1 l_1 \\ K_2 l_2 - K_1 l_1 & K_1 l_1^2 + K_2 l_2^2 - \omega^2 J_G \end{bmatrix} \begin{Bmatrix} A \\ \theta_0 \end{Bmatrix}$$

Eigen value sets $\Rightarrow \quad \omega_1 = 6.90rad/s = 1.10Hz$
$$\omega_2 = 9.06rad/s = 1.44Hz$$

$$\left(\frac{A}{\theta_0}\right)_{\omega_1} = -4.45m/rad$$

$$\left(\frac{A}{\theta_0}\right)_{\omega_2} = +0.33m/rad$$

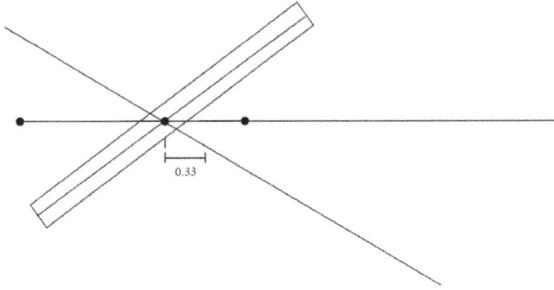

Practical Examples

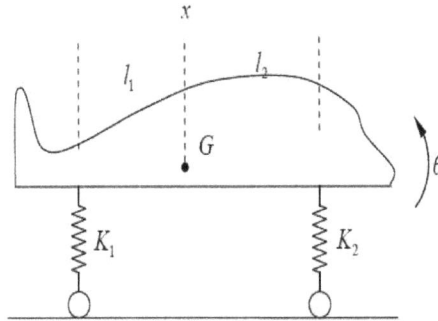

G= Centre of mass

K_1 = Total stiffness of a pair of springs representing the main loading gear.

K_2 = Stiffness of the spring of the main loading gear.

R= Radius of gyration of the a/c around the lateral axis through the centre of mass.

Same problem like a car

- restrained in translation and torsion

- Radius of gyration, r.

Forced vibration due to Harmonic excitation

$$\begin{bmatrix} m_{11} & m_{12} \\ m_{12} & m_{22} \end{bmatrix} \begin{Bmatrix} \ddot{x}_1 \\ \ddot{x}_2 \end{Bmatrix} + \begin{bmatrix} c_{11} & c_{12} \\ c_{12} & c_{22} \end{bmatrix} \begin{Bmatrix} \ddot{x}_1 \\ \ddot{x}_2 \end{Bmatrix} + \begin{bmatrix} k_{11} & k_{12} \\ k_{12} & k_{22} \end{bmatrix} \begin{Bmatrix} x_1 \\ x_2 \end{Bmatrix} = \begin{Bmatrix} F_1 \\ F_2 \end{Bmatrix} \sin \omega t$$

For steady state response,

$$x_1(t) = A_1 \sin \omega t$$
$$x_2(t) = A_2 \sin \omega t$$

Substituting above equations,

$$\begin{bmatrix} -\omega^2 m_{11} + i\omega c_{11} + k_{11} & -\omega^2 m_{12} + i\omega c_{12} + k_{12} \\ -\omega^2 m_{12} + i\omega c_{12} + k_{12} & -\omega^2 m_{22} + i\omega c_{22} + k_{22} \end{bmatrix} \begin{Bmatrix} A_1 \\ A_2 \end{Bmatrix} = \begin{Bmatrix} F_1 \\ F_2 \end{Bmatrix}$$
$$\Rightarrow [Z(\omega)]\{A\} = \{F\}$$
$$\uparrow$$

Impedence matrix

$$\{A\} = [Z(\omega)]^{-1}\{F\}$$

$$[Z(\omega)]^{-1} = \frac{Adj[Z(\omega)]}{|Z(\omega)|}$$

$$= \frac{1}{Z_{11}(\omega)Z_{22}(\omega) - Z_{12}^2(\omega)} \begin{bmatrix} Z_{22}(\omega) & -Z_{12}(\omega) \\ -Z_{12}(\omega) & Z_{11}(\omega) \end{bmatrix}$$

$$A_1(\omega) = \left[\frac{(k_{22} - \omega^2 m_2) F_1}{(k_{11} - \omega^2 m_1)(k_{22} - \omega^2 m_2) - k_{12}^2} \right]$$

$$F_2 = 0 \ and \ c_{ij} = 0$$

$$A_2(\omega) = \left[\frac{-k_{12} F_1}{(k_{11} - \omega^2 m_1)(k_{22} - \omega^2 m_2) - k_{12}^2} \right]$$

$$\begin{aligned} k_{11} &= k_1 + k_2 & m_{11} &= m_1 \\ k_{22} &= k_2 + k_3 & m_{22} &= m_2 \\ k_{12} &= -k_2 & m_{12} &= 0 \end{aligned}$$

Response of 2DOF System to Harmonic Excitation

$$m_{11}\ddot{x}_1 + m_{12}\ddot{x}_2 + c_{11}\dot{x}_1 + c_{12}\dot{x}_2 + K_{11}x_1 + K_{12}x_2 = F(t)$$

$$m_{12}\ddot{x}_2 + m_{22}\ddot{x}_2 + c_{12}\dot{x}_1 + c_{22}\dot{x}_2 + K_{12}x_1 + K_{22}x_2 = F_2(t)$$

Let us consider the following harmonic excitation

$$f_1(t) = F_1 e^{i\omega t} \qquad f_2(t) = F_2 e^{i\omega t}$$

and write the Steady-State response as

$$x_1(t) = X_1 e^{i\omega t} \qquad x_2(t) = X_2 e^{i\omega t}$$

Where X_1 and X_2 are in general complex quantities depending on the driving frequency ω and the system parameters. Inserting above we get the algebraic equations.

$$\left(-\omega^2 m_{11} + i\omega c_{11} + K_{11}\right) X_1 + \left(-\omega^2 m_{12} + i\omega c_{12} + K_{12}\right) X_2 = F_1$$

$$\left(-\omega^2 m_{12} + i\omega c_{12} + K_{12}\right) X_1 + \left(-\omega^2 m_{22} + i\omega c_{22} + K_{22}\right) X_2 = F_2$$

Introducing the notation

$$z_{ij}(\omega) = -\omega^2 m_{ij} + i\omega c_{ij} + K_{ij} \qquad i,j = 1,2$$

Where $z_{ij}(\omega)$ are known as impedances, Equations can be written in compact matrix form as

$$\left[z(\omega)\right]\{X\} = \{F\}$$

Where $\left[z(\omega)\right]$ is called as impedance matrix, $\{x\}$ is the column matrix of displacement amplitudes, and $\{F\}$ is the column matrix of the excitation amplitudes.

The solution can be obtained by pre-multiplying amplitudes of the equations by the inverse $\left[z(\omega)\right]^{-1}$ of the impedance matrix $\left[z(\omega)\right]$,

$$\{X\} = \left[Z(\omega)\right]^{-1}\{F\}$$

$$\left[Z(\omega)\right]^{-1} = \begin{bmatrix} Z_{22}(\omega) & -Z_{12}(\omega) \\ -Z_{12}(\omega) & Z_{11}(\omega) \end{bmatrix} \times \frac{1}{\det\left[Z(\omega)\right]}$$

$$= \frac{1}{Z_{11}(\omega)Z_{22}(\omega) - Z_{12}^2(\omega)} \begin{bmatrix} Z_{22}(\omega) & -Z_{12}(\omega) \\ -Z_{12}(\omega) & Z_{11}(\omega) \end{bmatrix}$$

Where

$$X_1(\omega) = \frac{Z_{22}(\omega)F_1 - Z_{12}(\omega)F_2}{Z_{11}(\omega)Z_{22}(\omega) - Z_{12}^2(\omega)}$$

$$X_2(\omega) = \frac{Z_{11}(\omega)F_2 - Z_{12}(\omega)F_1}{Z_{11}(\omega)Z_{22}(\omega) - Z_{12}^2(\omega)}$$

We note that $X_1(\omega)$ and $X_2(\omega)$ are frequency response take a system

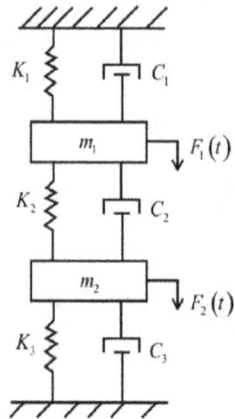

Simplified system

$$F_2 = 0 \quad c_{ij} = 0$$

Response

$$X_1(\omega) = \frac{\left(K_{22} - \omega^2 m_{22}\right)F_1}{\left(K_{11} - \omega^2 m_{11}\right)\left(K_{22} - \omega^2 m_{22}\right) - K_{12}^2}$$

$$X_2(\omega) = \frac{-K_{12}F_1}{\left(K_{11} - \omega^2 m_{11}\right)\left(K_{22} - \omega^2 m_{22}\right) - K_{12}^2}$$

Where

$$\left. \begin{array}{l} K_{11} = K_1 + K_2 \\ K_{22} = K_2 + K_3 \\ K_{12} = -K_2 \end{array} \right\} . \left. \begin{array}{l} m_{11} = m_1 \\ m_{22} = m_2 \\ m_{12} = 0 \end{array} \right.$$

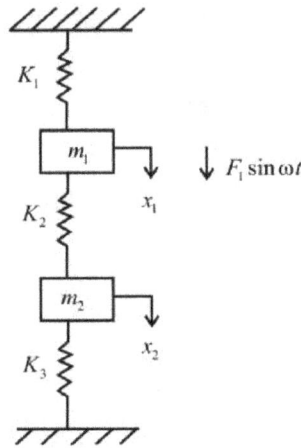

Assume

$$m_1 = m;\ m_2 = 2m$$

$$K_1 = K_2 = K;\ K_3 = 2K$$

$$\begin{aligned} K_{11} &= K_1 + K_2 = 2K \\ K_{22} &= K_2 + K_3 = 3K \\ K_{12} &= -K_2 = -K \end{aligned} \quad \begin{aligned} m_{11} &= m \\ m_{22} &= 2m \end{aligned}$$

$$X_1 = (\omega) = \frac{\left(3K - 2m\omega^2\right)F_1}{\left(2K - \omega^2 m\right)\left(3K - \omega^2 . 2m\right) - K^2}$$

$$= \frac{\left(3K - 2m\omega^2\right)F_1}{5K^2 - 7K\omega^2 m + 2\omega^4 m^2}$$

$$= \frac{2F_1}{3K} \cdot \left[\frac{\dfrac{3}{2} - \left(\dfrac{\omega}{\omega_1}\right)^2}{1 - \left(\dfrac{\omega}{\omega_1}\right)^2\left[1 - \left(\dfrac{\omega}{\omega_2}\right)^2\right]} \right]$$

$$\omega_1^2 = \frac{K}{m};\ \omega_2^2 = \frac{5}{2} . \frac{K}{m}$$

$$X_2(\omega) = \frac{KF_1}{2m^2\omega^4 - 7mK\omega^2 + 5K^2}$$

The denominator of X_1 and X_2 are recognized as the characteristic determinant, which can be written as,

$$\Delta(\omega^2) = 2m^2\omega^4 - 7mK\omega^2 + 5K^2$$

$$= 2m^2\omega^4 - 2mK\omega^2 - 5mK\omega^2 + 5K^2 = 2m^2\omega(m\omega^2 - K)5K(m\omega^2 - K)$$

$$= 2m^2(\omega^2 - \omega_1^2)(\omega^2 - \omega_2^2) = (m\omega^2 - K)(2m\omega^2 - 5K)$$

$$= 2m^2\left(\omega^2 - \frac{K}{m}\right)\left(\omega^2 - \frac{5}{2}\frac{K}{m}\right)$$

$$= 2m^2\left(\frac{K}{m}\right)\left\{1 - \left(\frac{\omega}{\omega_1}\right)^2\right\}\left(\frac{5}{2}\frac{K}{m}\right)\left(1 - \left(\frac{\omega}{\omega_1}\right)^2\right)$$

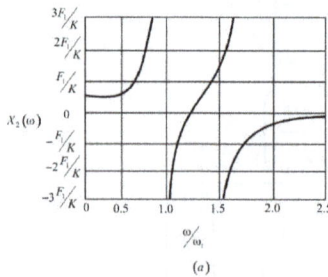

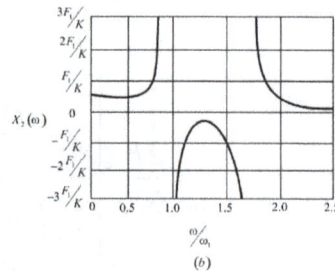

$$X_1(\omega) = \frac{2F_1}{5K}\frac{\frac{3}{2} - \left(\frac{\omega}{\omega_1}\right)^2}{\left[1 - (\omega/\omega_1)^2\right]\left[1 - (\omega/\omega_2)^2\right]}$$

$$X_2(\omega) = \frac{F_1}{5K}\frac{1}{\left[1 - \left(\frac{\omega}{\omega_1}\right)^2\right]\left[1 - (\omega/\omega_2)^2\right]}$$

(a)   (b)

Vibration Absorber

Let us consider the system m_1 and K_1 is the original SDOF system and the absorber consists of the mass m_2 and the spring K_2.

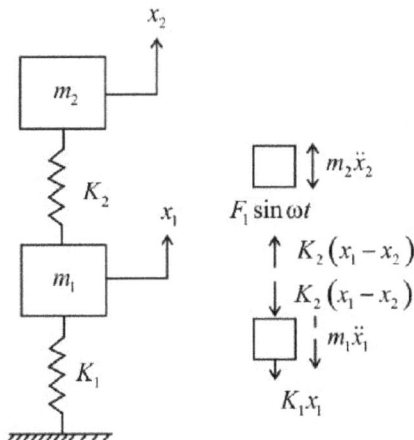

Equilibrium Equation

$$m_1\ddot{x}_1 + \left(K_1 + K_2\right)x_1 - K_2 x_2 = F_1 \sin \omega t$$

$$m_2\ddot{x}_2 - K_2 x_1 + K_2 x_2 = 0$$

$$m_2\ddot{x}_2 - K_2\left(x_1 - x_2\right) = 0 \Rightarrow m_2\ddot{x}_L - K_1 x_1 + K_2 x_2 = 0$$

$$m_1\ddot{x}_1 + K_1 x_1 + K_2\left(x_1 - x_2\right) = F_1 \sin \omega t$$

Let $x_1 = x_1 \sin \omega t \ x_2 = x_2 \sin \omega t$

We obtain two algebraic equations for $X_1 \ and \ X_2$ having matrix form,

$$\begin{bmatrix} K_1 + K_2 - \omega^2 m_1 & -K_2 \\ -K_2 & K_2 - \omega^2 m_2 \end{bmatrix} \begin{Bmatrix} X_1 \\ X_2 \end{Bmatrix} = \begin{Bmatrix} F_1 \\ 0 \end{Bmatrix}$$

$$X_1 - \frac{\left(K_2 - \omega^2 m_2\right)F_1}{\left(K_1 + K_2 - \omega^2 m_1\right)\left(K_2 - \omega^2 m_2\right) - K_2^2}$$

$$X_2 = \frac{K_2 F_1}{\left(K_1 + K_2 - \omega^2 m_1\right)\left(K_2 - \omega^2 m_2\right) - K_2^2}$$

$$X_1 = \frac{K_2\left(1 - \dfrac{\omega^2 m_2}{K_2}\right)F_1}{K_1\left(1 + \dfrac{K_2}{K_1} - \dfrac{\omega^2 m_1}{K_1}\right)K_2\left(1 - \dfrac{\omega^2 m_2}{K_2}\right) - K_1 K_2\left(\dfrac{K_2}{K_1}\right)}$$

$$= \frac{\left\{1-\left(\dfrac{\omega}{\omega_a}\right)^2\right\}F_1/K_1}{\left\{1+\mu.\dfrac{m_1}{m_2}.\dfrac{K_2}{K_1}-\left(\dfrac{\omega}{\omega_n}\right)^2\right\}\left\{1-\left(\dfrac{\omega}{\omega_a}\right)^2\right\}-\dfrac{m_2}{m_1}.\dfrac{m_1}{m_2}\dfrac{K_2}{K_1}}$$

$\omega_n = \sqrt{\dfrac{K_1}{m_1}} \rightarrow$ Natural frequency of main system

$\omega_\alpha = \sqrt{\dfrac{K_2}{m_2}} \rightarrow$ Natural frequency of absorber alone

$x_{st} = F_1/K_1 \rightarrow$ Static deflection of the main system

$\mu = \dfrac{m_2}{m_1} \rightarrow$ Ratio of the absorber mass to main mass

We can rewrite as

$$X_1 = \frac{\left[1-\left(\omega/\omega_\alpha\right)^2\right]x_{st}}{\left[1+\mu\left(\omega_\alpha/\omega\right)^2\right]\left[1-\left(\dfrac{\omega}{\omega_\alpha}\right)^2\right]-\mu\left(\dfrac{\omega}{\omega_\alpha}\right)^2}$$

If $\omega = \omega_\alpha$, the amplitude X_1 of the main mass reduce to zero.

$$X_2 = -\left(\dfrac{\omega_n}{\omega_\alpha}\right)^2 \dfrac{X_{st}}{\mu} = -\dfrac{F_1}{K_2}$$

So that, $= -\dfrac{m_1\omega_n^2 X_{st}}{m_2\omega_\alpha^2} = -\dfrac{F_1}{K_2}$

$$X_2(t) = -\dfrac{F_1}{K_2}\sin \omega t$$

Force in the absorber spring at any time is

$$K_2 x_2(t) = -F_1 \sin \omega t$$

Distributed Parameter System

A distributed parameter system (as opposed to a lumped parameter system) is a system whose state space is infinite-dimensional. Such systems are therefore also known as infinite-dimensional systems. Typical examples are systems described by partial differential equations or by delay differential equations.

Linear time-invariant Distributed Parameter Systems

Abstract Evolution Equations

Discrete-time

With U, X and Y Hilbert spaces and $A \in L(X)$, $B \in L(U, X)$, $C \in L(X, Y)$ and $D \in L(U, Y)$ the following equations determine a discrete-time linear time-invariant system:

$$x(k+1) = Ax(k) + Bu(k)$$

$$y(k) = Cx(k) + Du(k)$$

with x (the state) a sequence with values in X, u (the input or control) a sequence with values in U and y (the output) a sequence with values in Y.

Continuous-time

The continuous-time case is similar to the discrete-time case but now one considers differential equations instead of difference equations:

$$\dot{x}(t) = Ax(t) + Bu(t)_,$$

$$y(t) = Cx(t) + Du(t).$$

An added complication now however is that to include interesting physical examples such as partial differential equations and delay differential equations into this abstract framework, one is forced to consider unbounded operators. Usually A is assumed to generate a strongly continuous semigroup on the state space X. Assuming B, C and D to be bounded operators then already allows for the inclusion of many interesting physical examples, but the inclusion of many other interesting physical examples forces unboundedness of B and C as well.

Example: a Partial Differential Equation

The partial differential equation with $t > 0$ and $\xi \in [0, 1]$ given by

$$\frac{\partial}{\partial t} w(t, \xi) = -\frac{\partial}{\partial \xi} w(t, \xi) + u(t),$$

$$w(0,\xi) = w_0(\xi),$$

$$w(t,0) = 0,$$

$$y(t) = \int_0^1 w(t,\xi)d\xi,$$

fits into the abstract evolution equation framework described above as follows. The input space U and the output space Y are both chosen to be the set of complex numbers. The state space X is chosen to be $L^2(0, 1)$. The operator A is defined as

$$Ax = -x', \quad D(A) = \left\{x \in X : x \text{ absolutely continuous}, x' \in L^2(0,1) \text{ and } x(0) = 0\right\}.$$

It can be shown that A generates a strongly continuous semigroup on X. The bounded operators B, C and D are defined as

$$Bu = u, \quad Cx = \int_0^1 x(\xi)d\xi, \quad D = 0.$$

Example: a Delay Differential Equation

The delay differential equation

$$\dot{w}(t) = w(t) + w(t - \tau) + u(t),$$

$$y(t) = w(t),$$

fits into the abstract evolution equation framework described above as follows. The input space U and the output space Y are both chosen to be the set of complex numbers. The state space X is chosen to be the product of the complex numbers with $L^2(-\tau, 0)$. The operator A is defined as

$$A\begin{pmatrix} r \\ f \end{pmatrix} = \begin{pmatrix} r + f(-\tau) \\ f' \end{pmatrix}, \quad D(A) = \left\{ \begin{pmatrix} r \\ f \end{pmatrix} \in X : f \text{ absolutely continuous}, f' \in L^2([-\tau,0]) \text{ and } r = f(0) \right\}.$$

It can be shown that A generates a strongly continuous semigroup on X. The bounded operators B, C and D are defined as

$$Bu = \begin{pmatrix} u \\ 0 \end{pmatrix}, \quad C\begin{pmatrix} r \\ f \end{pmatrix} = r, \quad D = 0.$$

Transfer Functions

As in the finite-dimensional case the transfer function is defined through the Laplace transform (continuous-time) or Z-transform (discrete-time). Whereas in the finite-dimensional case the transfer function is a proper rational function, the infinite-dimensionality of the state space leads to irrational functions (which are however still holomorphic).

Discrete-time

In discrete-time the transfer function is given in terms of the state space parameters by $D + \sum_{k=0}^{\infty} CA^k Bz^k$ and it is holomorphic in a disc centered at the origin. In case $1/z$ belongs to the resolvent set of A (which is the case on a possibly smaller disc centered at the origin) the transfer function equals $D + Cz(I - zA)^{-1}B$. An interesting fact is that any function that is holomorphic in zero is the transfer function of some discrete-time system.

Continuous-time

If A generates a strongly continuous semigroup and B, C and D are bounded operators, then the transfer function is given in terms of the state space parameters by $D + C(sI - A)^{-1}B$ for s with real part larger than the exponential growth bound of the semigroup generated by A. In more general situations this formula as it stands may not even make sense, but an appropriate generalization of this formula still holds. To obtain an easy expression for the transfer function it is often better to take the Laplace transform in the given differential equation than to use the state space formulas as illustrated below on the examples given above.

Transfer Function for the Partial Differential Equation Example

Setting the initial condition w_0 equal to zero and denoting Laplace transforms with respect to t by capital letters we obtain from the partial differential equation given above

$$sW(s,\xi) = -\frac{d}{d\xi}W(s,\xi) + U(s),$$

$$W(s,0) = 0,$$

$$Y(s) = \int_0^1 W(s,\xi)d\xi.$$

This is an inhomogeneous linear differential equation with ξ as the variable, s as a parameter and initial condition zero. The solution is $W(s,\xi)=U(s)(1-e^{-s\xi})/s$. Substituting this in the equation for Y and integrating gives $Y(s)=U(s)(e^{-s}+s-1)/s^2$ so that the transfer function is $(e^{-s}+s-1)/s^2$.

Transfer Function for the Delay Differential Equation Example

Proceeding similarly as for the partial differential equation example, the transfer function for the delay equation example is $1/\left(s-1-e^{-s}\right)$.

Controllability

In the infinite-dimensional case there are several non-equivalent definitions of controllability which for the finite-dimensional case collapse to the one usual notion of controllability. The three most important controllability concepts are:

- Exact controllability,

- Approximate controllability,

- Null controllability.

Controllability in Discrete-time

An important role is played by the maps Φ_n which map the set of all U valued sequences into X and are given by $\Phi_n u = \sum_{k=0}^{n} A^k B u_k$. The interpretation is that $\Phi_n u$ is the state that is reached by applying the input sequence u when the initial condition is zero. The system is called

- exactly controllable in time n if the range of Φ_n equals X,

- approximately controllable in time n if the range of Φ_n is dense in X,

- null controllable in time n if the range of Φ_n includes the range of A^n.

Controllability in Continuous-time

In controllability of continuous-time systems the map Φ_t given by $\int_0^t e^{As} Bu(s)ds$ plays the role that Φ_n plays in discrete-time. However, the space of control functions on which this operator acts now influences the definition. The usual choice is $L^2(0,\infty;U)$, the space of (equivalence classes of) U-valued square integrable functions on the interval $(0,\infty)$, but other choices such as $L^1(0,\infty;U)$ are possible. The different controllability notions can be defined once the domain of Φ_t is chosen. The system is called

- exactly controllable in time t if the range of Φ_t equals X,

- approximately controllable in time t if the range of Φ_t is dense in X,

- null controllable in time t if the range of Φ_t includes the range of e^{At}.

Observability

As in the finite-dimensional case, observability is the dual notion of controllability. In the infinite-dimensional case there are several different notions of observability which in the finite-dimensional case coincide. The three most important ones are:

- Exact observability (also known as continuous observability),

- Approximate observability,

- Final state observability.

Observability in Discrete-time

An important role is played by the maps Ψ_n which map X into the space of all Y valued sequences and are given by $(\Psi_n x)_k = CA^k x$ if $k \le n$ and zero if $k > n$. The interpretation is that $\Psi_n x$ is the truncated output with initial condition x and control zero. The system is called:

- exactly observable in time n if there exists a $k_n > 0$ such that $\| \Psi_n x \| \ge k_n \| x \|$ for all $x \in X$,

- approximately observable in time n if Ψ_n is injective,

- final state observable in time n if there exists a $k_n > 0$ such that $\| \Psi_n x \| \ge k_n \| A^n x \|$ for all $x \in X$.

Observability in Continuous-time

In observability of continuous-time systems the map Ψ_t given by $(\Psi_t)(s) = Ce^{As}x$ for $s \in [0,t]$ and zero for $s > t$ plays the role that Ψ_n plays in discrete-time. However, the space of functions to which this operator maps now influences the definition. The usual choice is $L^2(0, \infty, Y)$, the space of (equivalence classes of) Y-valued square integrable functions on the interval $(0,\infty)$, but other choices such as $L^1(0, \infty, Y)$ are possible. The different observability notions can be defined once the co-domain of Ψ_t is chosen. The system is called:

- exactly observable in time t if there exists a $k_t > 0$ such that $\| \Psi_t x \| \ge k_t \| x \|$ for all $x \in X$,

- approximately observable in time t if Ψ_t is injective,

- final state observable in time t if there exists a $k_t > 0$ such that $\| \Psi_t x \| \geq k_t \| e^{At} x \|$ for all $x \in X$.

Duality Between Controllability and Observability

As in the finite-dimensional case, controllability and observability are dual concepts (at least when for the domain of Φ and the co-domain of Ψ the usual L^2 choice is made). The correspondence under duality of the different concepts is:

- Exact controllability $\leftrightarrow$ Exact observability,

- Approximate controllability $\leftrightarrow$ Approximate observability,

- Null controllability $\leftrightarrow$ Final state observability.

Ritz Method

The Ritz method is a direct method to find an approximate solution for boundary value problems. The method is named after Walter Ritz.

In quantum mechanics, a system of particles can be described in terms of an "energy functional" or Hamiltonian, which will measure the energy of any proposed configuration of said particles. It turns out that certain privileged configurations are more likely than other configurations, and this has to do with the eigenanalysis ("analysis of characteristics") of this Hamiltonian system. Because it is often impossible to analyze all of the infinite configurations of particles to find the one with the least amount of energy, it becomes essential to be able to approximate this Hamiltonian in some way for the purpose of numerical computations.

The Ritz method can be used to achieve this goal. In the language of mathematics, it is exactly the finite element method used to compute the eigenvectors and eigenvalues of a Hamiltonian system.

Discussion

As with other variational methods, a trial wave function, Ψ, is tested on the system. This trial function is selected to meet boundary conditions (and any other physical constraints). The exact function is not known; the trial function contains one or more adjustable parameters, which are varied to find a lowest energy configuration.

It can be shown that the ground state energy, E_0, satisfies an inequality:

$$E_0 \le \frac{\langle \Psi \mid \hat{H} \mid \Psi \rangle}{\langle \Psi \mid \Psi \rangle}.$$

That is, the ground-state energy is less than this value. The trial wave-function will always give an expectation value larger than or equal to the ground-energy.

If the trial wave function is known to be orthogonal to the ground state, then it will provide a boundary for the energy of some excited state.

The Ritz ansatz function is a linear combination of N known basis functions $\{\Psi_i\}$, parametrized by unknown coefficients:

$$\Psi = \sum_{i=1}^{N} c_i \Psi_i.$$

With a known Hamiltonian, we can write its expected value as

$$\varepsilon = \frac{\left\langle \sum_{i=1}^{N} c_i \Psi_i \mid \hat{H} \mid \sum_{i=1}^{N} c_i \Psi_i \right\rangle}{\left\langle \sum_{i=1}^{N} c_i \Psi_i \mid \sum_{i=1}^{N} c_i \Psi_i \right\rangle} = \frac{\sum_{i=1}^{N} \sum_{j=1}^{N} c_i^* c_j H_{ij}}{\sum_{i=1}^{N} \sum_{j=1}^{N} c_i^* c_j S_{ij}} \equiv \frac{A}{B}.$$

The basis functions are usually not orthogonal, so that the overlap matrix S has nonzero nondiagonal elements. Either $\{c_i\}$ or $\{c_i^*\}$ (the conjugation of the first) can be used to minimize the expectation value. For instance, by making the partial derivatives of ε over $\{c_i^*\}$ zero, the following equality is obtained for every $k = 1, 2, ..., N$:

$$\frac{\partial \varepsilon}{\partial c_k^*} = \frac{\sum_{j=1}^{N} c_j (H_{kj} - \varepsilon S_{kj})}{B} = 0,$$

which leads to a set of N secular equations:

$$\sum_{j=1}^{N} c_j \left(H_{kj} - \varepsilon S_{kj} \right) = 0 \quad for \quad k = 1, 2, ..., N.$$

In the above equations, energy ε and the coefficients $\{c_j\}$ are unknown. With respect to $\mathbf{c}$, this is a homogeneous set of linear equations, which has a solution when the determinant of the coefficients to these unknowns is zero:

$$\det\left(H - \varepsilon S \right) = 0,$$

which in turn is true only for N values of ε. Furthermore, since the Hamiltonian is a hermitian operator, the H matrix is also hermitian and the values of ε_i will be real. The lowest value among ε_i (i=1,2,..,N), ε_0, will be the best approximation to the ground state for the basis functions used. The remaining N-1 energies are estimates of excited state energies. An approximation for the wave function of state i can be obtained by finding the coefficients $\left\{c_j\right\}$ from the corresponding secular equation.

The Relationship with the Finite Element Method

In the language of the finite element method, the matrix H_{kj} is precisely the *stiffness matrix* of the Hamiltonian in the piecewise linear element space, and the matrix S_{kj} is the *mass matrix*. In the language of linear algebra, the value ϵ is an eigenvalue of the discretized Hamiltonian, and the vector c is a discretized eigenvector.

Galerkin Method

In mathematics, in the area of numerical analysis, Galerkin methods are a class of methods for converting a continuous operator problem (such as a differential equation) to a discrete problem. In principle, it is the equivalent of applying the method of variation of parameters to a function space, by converting the equation to a weak formulation. Typically one then applies some constraints on the function space to characterize the space with a finite set of basis functions. Galerkin's method provides powerful numerical solution to differential equations and modal analysis.

The approach is usually credited to Boris Galerkin but the method was discovered by Walther Ritz, to whom Galerkin refers. Often when referring to a Galerkin method, one also gives the name along with typical approximation methods used, such as Bubnov–Galerkin method (after Ivan Bubnov), Petrov–Galerkin method (after Georgii I. Petrov) or Ritz–Galerkin method (after Walther Ritz).

Examples of Galerkin methods are:

- the Galerkin method of weighted residuals, the most common method of calculating the global stiffness matrix in the finite element method,

- the boundary element method for solving integral equations,

- Krylov subspace methods.

Introduction with an Abstract Problem

A Problem in Weak Formulation

Let us introduce Galerkin's method with an abstract problem posed as a weak formulation on a Hilbert space V , namely, find $u \in V$ such that for all $v \in V, a(u,v) = f(v)$.

Here, $a(\cdot,\cdot)$ is a bilinear form (the exact requirements on $a(\cdot,\cdot)$ will be specified later) and f is a bounded linear functional on V.

Galerkin Dimension Reduction

Choose a subspace $V_n \subset V$ of dimension n and solve the projected problem:

Find $u_n \in V_n$ such that for all $v_n \in V_n, a(u_n, v_n) = f(v_n)$.

We call this the Galerkin equation. Notice that the equation has remained unchanged and only the spaces have changed. Reducing the problem to a finite-dimensional vector subspace allows us to numerically compute u_n as a finite linear combination of the basis vectors in V_n.

Galerkin Orthogonality

The key property of the Galerkin approach is that the error is orthogonal to the chosen subspaces. Since $V_n \subset V$, we can use v_n as a test vector in the original equation. Subtracting the two, we get the Galerkin orthogonality relation for the error, $\epsilon_n = u - u_n$ which is the error between the solution of the original problem, u, and the solution of the Galerkin equation, u_n

$$a(\epsilon_n, v_n) = a(u, v_n) - a(u_n, v_n) = f(v_n) - f(v_n) = 0.$$

Matrix Form

Since the aim of Galerkin's method is the production of a linear system of equations, we build its matrix form, which can be used to compute the solution by a computer program.

Let $e_1, e_2, \ldots, e_n$ be a basis for V_n. Then, it is sufficient to use these in turn for testing the Galerkin equation, i.e.: find $u_n \in V_n$ such that

$$a(u_n, e_i) = f(e_i) \quad i = 1, \ldots, n.$$

We expand u_n with respect to this basis, $u_n = \sum_{j=1}^{n} u_j e_j$ and insert it into the equation above, to obtain

$$a\left(\sum_{j=1}^{n} u_j e_j, e_i \right) = \sum_{j=1}^{n} u_j a(e_j, e_i) = f(e_i) \quad i = 1, \ldots, n.$$

This previous equation is actually a linear system of equations $Au = f$, where

$$A_{ij} = a(e_j, e_i), \quad f_i = f(e_i).$$

Symmetry of the Matrix

Due to the definition of the matrix entries, the matrix of the Galerkin equation is symmetric if and only if the bilinear form $a(\cdot, \cdot)$ is symmetric.

Analysis of Galerkin Methods

Here, we will restrict ourselves to symmetric bilinear forms, that is

$$a(u, v) = a(v, u).$$

While this is not really a restriction of Galerkin methods, the application of the standard theory becomes much simpler. Furthermore, a Petrov–Galerkin method may be required in the nonsymmetric case.

The analysis of these methods proceeds in two steps. First, we will show that the Galerkin equation is a well-posed problem in the sense of Hadamard and therefore admits a unique solution. In the second step, we study the quality of approximation of the Galerkin solution u_n.

The analysis will mostly rest on two properties of the bilinear form, namely

- Boundedness: for all $u, v \in V$ holds

  $a(u, v) \le C \|u\| \|v\|$ for some constant $C > 0$

- Ellipticity: for all $u \in V$ holds

  $a(u, u) \ge c \|u\|^2$ for some constant $c > 0$.

By the Lax-Milgram theorem, these two conditions imply well-posedness of the original problem in weak formulation.

Well-posedness of the Galerkin Equation

Since $V_n \subset V$, boundedness and ellipticity of the bilinear form apply to V_n. Therefore, the well-posedness of the Galerkin problem is actually inherited from the well-posedness of the original problem.

Quasi-best Approximation (Céa's Lemma)

The error $u - u_n$ between the original and the Galerkin solution admits the estimate

$$\|u - u_n\| \le \frac{C}{c} \inf_{v_n \in V_n} \|u - v_n\|.$$

This means, that up to the constant C/c, the Galerkin solution u_n is as close to the original solution u as any other vector in V_n. In particular, it will be sufficient to study approximation by spaces V_n, completely forgetting about the equation being solved.

Proof

Since the proof is very simple and the basic principle behind all Galerkin methods, we include it here: by ellipticity and boundedness of the bilinear form (inequalities) and Galerkin orthogonality (equals sign in the middle), we have for arbitrary $v_n \in V_n$:

$$c \| u - u_n \|^2 \le a(u - u_n, u - u_n) = a(u - u_n, u - v_n) \le C \| u - u_n \| \| u - v_n \|.$$

Dividing by $c \| u - u_n \|$ and taking the infimum over all possible v_n yields the lemma.

Steps of Modal Analysis

1) $\{d_1\}$ and $\{d_2\} \to$ orthogonal if $d_1^T d_2 = 0$

2) Norm $\|d\| = \sqrt{d^T d} = \left[\sum_{i=1}^{n} (d_i)^2 \right]^{1/2}$

3) $\|d\| = 1$ orthogonal

4) $[\Lambda] = dia[\lambda_i] = [d^T][\bar{K}][d]$

5) $\{d\}$ is normalised by $\dfrac{1}{\sqrt{d^T d}} . d$

Cholesky Decomposition

When matrix $[M]$ or $[K]$ is full populated, matrices $[L]$ and $[L]^T$ can be found from Cholesky decomposition. In this evaluation, we simply write the $[M] = [L][L]^T$ in terms of the upper and lower triangular matrix of $[L]$ and its transpose.

$$[M]\{\ddot{x}\} + [K]\{x\} = 0$$

- Coupling comes from K

Take $[L]$ - matrix in such a way that

$$[M]=[L][L]^T$$

$$\therefore L = [M]^{1/2} = \begin{bmatrix} \sqrt{m} & 0 & 0 \\ 0 & . & 0 \\ 0 & 0 & \sqrt{m} \end{bmatrix}$$

$$[M]=[L][L]^T$$

$$[L]^{-1} = [M]^{\frac{1}{2}} = \begin{bmatrix} \dfrac{1}{\sqrt{m}} & 0 \\ 0 & \dfrac{1}{\sqrt{m}} \end{bmatrix}$$

$$[M]\{\ddot{x}\}+[K]\{x\}=0$$

$$\{x\}=[M]^{-\frac{1}{2}}\{q\}$$

Sustituting $\{x\}$ and pre-multiplying by $[M]^{-\frac{1}{2}}$ to the above equation, we can rewrite

$$[M]^{-\frac{1}{2}}[M][M]^{-\frac{1}{2}}\{q\}+[M]^{-\frac{1}{2}}[K][M]^{-\frac{1}{2}}\{q\}=0$$

$$\Rightarrow [I]\{\ddot{q}\}+\left[\bar{K}\right]\{q\}$$

Now defining

$$[q]=\{d\}e^{j\omega t}$$

Substituting above equations we get

$$\left[\bar{K}\right]\{d\}=\omega^2\{d\}$$
$$\Rightarrow \left[\left[\bar{K}\right]-\lambda[I]\right]\{d\}=0$$

$$\uparrow \qquad\qquad \nwarrow$$

Eigen values   *Eigen vector*

Continuous System

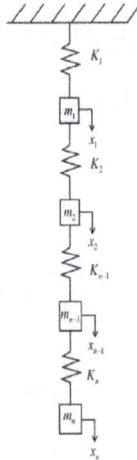

Free Undamped Vibration

$$m_i \ddot{x}_i + K_i \left(x_i - x_{i-1} \right) - K_{i+1} \left(x_{i+1} - x_i \right) = 0$$

Where $i = 1, 2, ..., n$

$$[M]\{\ddot{x}\} + [K]\{x\} = \{0\}$$

$$[K] = \begin{bmatrix} K_1 + K_2 & -K_2 & 0 & 0 \\ -K_2 & K_2 + K_3 & -K_3 & 0 \\ 0 & -K_3 & K_3 + K_4 & -K_4 \\ . & . & . & . \\ . & . & . & . \\ . & . & . & . \end{bmatrix} \quad [M] = \begin{bmatrix} m_1 & 0 & 0 & 0 \\ 0 & m_2 & 0 & 0 \\ 0 & 0 & m_3 & 0 \\ 0 & 0 & 0 & m_4 \\ . & . & . & . \\ . & . & . & . \end{bmatrix}$$

Modal Analysis

$$[M]\{\ddot{x}\} + [K]\{x\} = \{0\}$$

With initial condition $\{x(\ )\} \quad \{x \}$

$$\{\dot{x}(0)\} = \{\dot{x}_0\}$$

$\{x\} = [M]^{-\frac{1}{2}} \{q\}$ and substitute in (1)

$$[M][M]^{-\frac{1}{2}} \{q\} + [K][M]^{-\frac{1}{2}} \{q\} = 0$$

Premultiply by $[M]^{-\frac{1}{2}}$

$$[M]^{-\frac{1}{2}}[M][M]^{-\frac{1}{2}}\{\ddot{q}\}+[M]^{-\frac{1}{2}}[K][M]^{-\frac{1}{2}}\{q\}=0$$

$$\text{Or, } [I]\{\ddot{q}\}+[K]\{q\}=0$$

$\{q\}=[M]^{-\frac{1}{2}}\{x(t)\}$_____modal solution

Apply another transformation

$$\{q\}=[d]\{r\}$$

Where, $[d]=[d_1,d_2,....]d_i$'s are normalized vectors of $[\bar{K}]$

Substitute above equations and premultiply by $[d]^T$

$$[d]^T[d]\{\ddot{r}\}+[d]^T[\bar{K}][d]\{r\}=0$$

$$[I]\{\ddot{r}\}+[\Lambda]\{r\}=0$$

$$[\Lambda]=diag[\lambda_i]=diag[\omega_i^2]$$

From SDOF System

$$m\ddot{x}+Kx=0$$

$$x(0)=x_0 \quad \dot{x}(0)=\upsilon_0$$

$$x=\left[\left(\frac{\upsilon_0}{\omega_n}\right)^2+(x_0)^2\right]^{\frac{1}{2}}\sin(\omega_n t+\alpha)$$

$$\alpha=\tan^{-1}\frac{\omega_n x_0}{\upsilon_0}$$

Similarly modal response for first two modes:

$$r_1(t)=\sqrt{\frac{r_{10}^2\omega_{n1}^2+r_{10}^2}{\omega_{n1}}}\sin\left(\omega_{n1}'+\tan^{-1}\frac{\omega_{n1}r_{10}}{r_{10}}\right)$$

$$r_2(t) = \sqrt{\frac{r_{20}^2 \omega_{n2}^2 + r_{20}^2}{\omega_{n2}}} \sin\left(\omega_{n2}^t + \tan^{-1}\frac{\omega_{n2} r_{20}}{r_{20}}\right)$$

$$r_n(t) = \sqrt{\frac{r_{n0}^2 \omega_{nn}^2 + r_{n0}^2}{\omega_{nn}}} \sin\left(\omega_{nn}^t + \tan^{-1}\frac{\omega_{nn} r_{n0}}{r_{n0}}\right)$$

We assumed:

$$x = [M]^{-\frac{1}{2}}\{q\} = [M]^{-\frac{1}{2}}[d]\{r\}$$
$$= [S]\{r\}$$
$$\nearrow$$

matrix of mod *e shape vector*

The modal analysis can be easily extended to equations for a n dof system and easily a computer program may be written.

Calculation Steps:

1) Calculate $[M]^{-\frac{1}{2}}$

2) Calculate $\bar{K} = [M]^{-\frac{1}{2}}[K][M]^{-\frac{1}{2}}$

3) Obtain ω_i^2 and d_i

4) Normalize $[d_i]$ to form $[d]dd^T = 1$

5) Calculate $[S] = [M]^{-\frac{1}{2}}[d] \Rightarrow [S]^{-1} = [d]^T[M]^{\frac{1}{2}}$

6) Calculate the initial Coordinate in 'r' Co-ordinate(modal) systems

$$r_0 = [S]^{-1}\{x_0\} ; \dot{r}(0) = [S]^{-1}\{\dot{x}_0\}$$

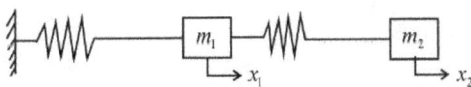

Physical Co-ordinate System

$$\{r\} = [S]^{-1}\{x\}$$

$$\Rightarrow x = [S]\{r\}$$

Where $[S] = [M]^{-\frac{1}{2}}[d]$

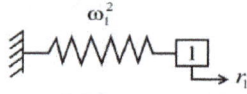

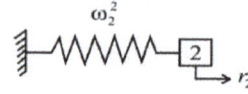

modal Co-ordinate (Uncoupled)

Example

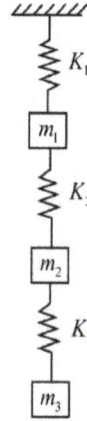

$$K_1 = K_2 = K_3 = 4 \, \frac{N}{m}$$

$$m_1 = m_2 = m_3 = 4Kg$$

Initial Condition

$$x_1(0) = 1$$

All other disp. and velocities are zero.

$$M = \begin{bmatrix} 4 & 0 & 0 \\ 0 & 4 & 0 \\ 0 & 0 & 4 \end{bmatrix} = 4[I]; \; [M]^{-\frac{1}{2}} = \frac{1}{2}[I]$$

$$K = \begin{bmatrix} 8 & -4 & 0 \\ -4 & 8 & -4 \\ 0 & -4 & 4 \end{bmatrix}$$

$$[\bar{K}] = [M]^{-\frac{1}{2}}[K][M]^{-\frac{1}{2}} = \frac{1}{4}\begin{bmatrix} 8 & -4 & 0 \\ -4 & 8 & -4 \\ 0 & -4 & 4 \end{bmatrix} = \begin{bmatrix} 2 & -1 & 0 \\ -1 & 2 & -1 \\ 0 & -1 & 1 \end{bmatrix}$$

$$\det\left(\lambda I - [\bar{K}]\right) = 0 \quad \det\left(\begin{bmatrix} \lambda-2 & 1 & 0 \\ 1 & \lambda-2 & 1 \\ 0 & 1 & \lambda-1 \end{bmatrix}\right) = 0$$

$$\lambda^3 - 5\lambda^2 + 6\lambda - 1 = 0 \Rightarrow (\lambda-2)\{(\lambda-2)(\lambda-1)-1\} - 1(-\lambda........)$$

$$\lambda_1 = 0.1981\, \omega_1 = 0.445 \Rightarrow (\lambda-2)(\lambda^2 - 3\lambda + 2 - 1) - (\lambda....)$$

$$\lambda_2 = 1.555\, \omega_2 = 1.247 \Rightarrow \lambda^3 - 3\lambda^2 + \lambda - 2\lambda^2 + 6\lambda - ....$$

$$\lambda_3 = 3.247\, \omega_3 = 1.8019 \Rightarrow \lambda^3 - 5\lambda^2 + 6\lambda - 1 = ...$$

Eigen Vector Calculation $\lambda_1 = 0.1981$

$$\left[\bar{K} - \lambda(I)\right]\{d_1\} = 0$$

Where $d_1 = \begin{bmatrix} d_{11} & d_{21} & d_{31} \end{bmatrix}^T$

$$\begin{bmatrix} 2.0-0.1981 & -1 & 0 \\ -1 & 2-0.1981 & -1 \\ 0 & -1 & 1-0.1981 \end{bmatrix}\begin{Bmatrix} d_{11} \\ d_{21} \\ d_{31} \end{Bmatrix} = \begin{bmatrix} 0 \\ 0 \\ 0 \end{bmatrix}$$

$$1.8019d_{11} - d_{21} = 0 \quad d_{11} = \frac{1}{1.8019}d_{21}$$

$$-d_{11} + 1.8019d_{21} - d_{31} = 0$$

$$-d_{21} + 0.8019d_{31} = 0 \quad d_{21} = 0.8019d_{31}$$

Substituting above equations

$$d_{11} = \frac{1}{1.8019} * (0.8019d_{31}) = 0.4450d_{31}$$

$$d_{21} = 0.8019d_{31}$$

$$d_1 = \begin{Bmatrix} d_{11} \\ d_{21} \\ d_{31} \end{Bmatrix} = d_{31} \begin{Bmatrix} 0.4450 \\ 0.8019 \\ 1 \end{Bmatrix}$$

Normalized $d_1^T . d_1 = 1$

$$d_1^T . d_1 = d_{31}^2 \left[(0.4450)^2 + (0.8019)^2 + 1^2 \right]$$

$$d_{31}^2 . [1.84106861] \qquad d_{31} = 1$$

$$\sqrt{d_1^T d_1} = \sqrt{1.84106861} = 1.3568598$$

$\therefore$ Normalized,

$$d_1 = \frac{1}{\sqrt{d^T d_1}} \begin{Bmatrix} d_{11} \\ d_{21} \\ d_{31} \end{Bmatrix} = \frac{1}{1.3568598} \begin{Bmatrix} 0.4450 \\ 0.8019 \\ 1 \end{Bmatrix}$$

$$= \begin{Bmatrix} 0.328 \\ 0.5910 \\ 0.7370 \end{Bmatrix}$$

Similarly, Normalized Vectors

$$\{d_2\} = \begin{Bmatrix} -0.7370 \\ -0.3280 \\ -0.5910 \end{Bmatrix} \qquad \{d_3\} = \begin{Bmatrix} -0.5910 \\ 0.7370 \\ -0.3280 \end{Bmatrix}$$

$$\{d\} = \begin{bmatrix} 0.3280 & -0.7370 & -0.5910 \\ 0.5910 & -0.3280 & +0.7370 \\ 0.7370 & -0.5910 & -0.3280 \end{bmatrix}$$

$$[d]^T [d] = [I]$$

$$[d]^K [\bar{K}][d] = [\Lambda]$$

$$[S] = [M]^{-\frac{1}{2}} [d] = \frac{1}{2} [I][d]$$

$$= \begin{bmatrix} 0.164 & -0.3685 & -0.2955 \\ 0.2955 & -0.1640 & 0.3685 \\ 0.3685 & -0.2955 & -0.1640 \end{bmatrix}$$

$$j^{-1} = [d]^T [M]^{-\frac{1}{2}} = 2[d]^T [I]$$

$$[S]^{-1} [S] = [I]$$

$$[S]^{-1} = \begin{bmatrix} 0.6560 & 1.1820 & 1.4740 \\ -1.4740 & -0.6560 & 1.1820 \\ -1.1820 & 1.4740 & -0.6560 \end{bmatrix}$$

Initial Conditions in the Modal Coordinates

$$\{r_0\} = [S]^{-1} \{\dot{x}_0\} = [S]^{-1} \{0\} = \{0\}$$

$$\{r(0)\} = [S]^{-1} \{\bar{x}_0\}$$

$$= \begin{bmatrix} 0.6560 & 1.1820 & 1.4740 \\ -1.4740 & -0.6560 & 1.1820 \\ -1.1820 & 1.4740 & -0.6560 \end{bmatrix} \begin{Bmatrix} 1 \\ 0 \\ 0 \end{Bmatrix}$$

$$= \begin{Bmatrix} 0.6560 \\ -1.4740 \\ -1.1820 \end{Bmatrix}$$

Modal Solution

$$r_1(t) = 0.6560 \operatorname{Sin}\left(0.445 + \frac{\pi}{2}\right) = 0.6560 \operatorname{Cos}(0.4457)$$

$$r_2(t) = (-1.4740) \operatorname{Sin}\left(1.247t + \frac{\pi}{2}\right) = -1.474 \operatorname{Cos}(1.247t)$$

$$r_3(t) = (-1.1820) \operatorname{Sin}\left(1.8019t + \frac{\pi}{2}\right) = -1.1820 \operatorname{Cos}(1.8019t)$$

Solution in physical Coordinates

$$\{x\} = [S]\{r(t)\}$$

$$= \begin{bmatrix} 0.164 & -0.3685 & -0.2955 \\ 0.2955 & -0.1640 & +0.3685 \\ 0.3685 & 0.2955 & -0.1640 \end{bmatrix} \begin{Bmatrix} 0.656\,\mathrm{Cos}(0.445t) \\ -1.474\,\mathrm{Cos}(1.247t) \\ -1.1820\,\mathrm{Cos}(1.8019t) \end{Bmatrix}$$

$$= \begin{Bmatrix} 0.1076\cos(0.445t) + 0.5432\cos(1.24t) + 0.3493\cos(1.8019t) \\ 0.1938\cos(0.445t) + 0.2417\cos(1.24t) + 0.4355\cos(1.8019t) \\ 0.2417\cos(0.445t) - 0.4355\cos(1.24t) + 0.1944\cos(1.8019t) \end{Bmatrix}$$

A problem is considered as two masses are connected a spring as shown below

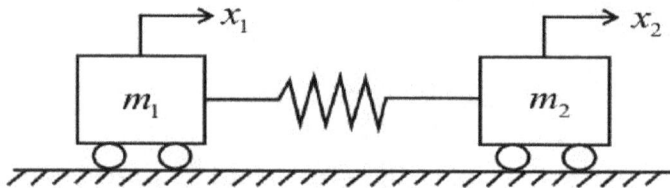

Rigid Body Mode:

-   Physical motion not involving the oscillation

The vibrating system may be an unrestrained system and may have an additional rigid body motion in the form of translation or rotation

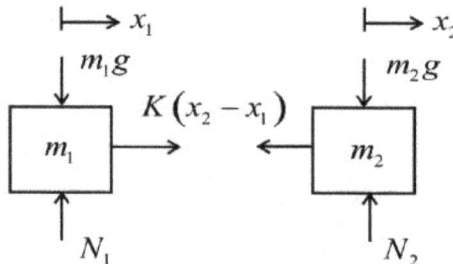

Initial Conditions

$$x(0) = \begin{Bmatrix} 0.01 \\ 0 \end{Bmatrix}$$

$$\dot{x}(0) = \begin{Bmatrix} 0 \\ 0 \end{Bmatrix}$$

$$m_1 \ddot{x}_1 = K(x_2 - x_1)$$
$$m_2 \ddot{x}_2 = -K(x_2 - x_1)$$

$$\begin{bmatrix} m_1 & 0 \\ 0 & m_2 \end{bmatrix} \begin{Bmatrix} \ddot{x}_1 \\ \ddot{x}_2 \end{Bmatrix} + K \begin{bmatrix} 1 & -1 \\ -1 & 1 \end{bmatrix} \begin{Bmatrix} x_1 \\ x_2 \end{Bmatrix} = \begin{Bmatrix} 0 \\ 0 \end{Bmatrix}$$

$$[M]^{-\frac{1}{2}} = \begin{bmatrix} 1 & 0 \\ 0 & \frac{1}{2} \end{bmatrix} \qquad M = \begin{bmatrix} 1 & 0 \\ 0 & 4 \end{bmatrix} \quad K = \begin{bmatrix} 4 & -4 \\ -4 & 4 \end{bmatrix}$$

$$[K] = [M]^{\frac{1}{2}} [K][M]^{-\frac{1}{2}}$$

$$= 4 \begin{bmatrix} 1 & 0 \\ 0 & \frac{1}{2} \end{bmatrix} \begin{bmatrix} 1 & -1 \\ -1 & 1 \end{bmatrix} \begin{bmatrix} 1 & 0 \\ 0 & \frac{1}{2} \end{bmatrix} = \begin{bmatrix} 4 & -2 \\ -2 & 1 \end{bmatrix}$$

$$\det([K] - \lambda[I]) = \begin{bmatrix} 4 - \lambda & -2 \\ -2 & 1 - \lambda \end{bmatrix}$$

$$\Rightarrow (\lambda^2 - 5\lambda) = 0$$

$$\lambda_1 = 0 \quad \lambda_2 = 5$$

$$\omega_{n1} = 0 \; ; \; \omega_{n2} = 2.235 \, rad \, / \, s$$

For $\lambda_1 = 0$

$$\begin{bmatrix} 4 - 0 & -2 \\ -2 & 1 - 0 \end{bmatrix} \begin{Bmatrix} d_{11} \\ d_{21} \end{Bmatrix} = \begin{Bmatrix} 0 \\ 0 \end{Bmatrix}$$

$$4d_{11} - 2d_{21} = 0 \Rightarrow$$

$$\begin{Bmatrix} d_{11} \\ d_{21} \end{Bmatrix} = \{d_1\} = \begin{Bmatrix} 1 \\ 2 \end{Bmatrix} \qquad d^T d = 1$$

For $\lambda_2 = 5\{d_2\} = \begin{Bmatrix} -2 \\ 1 \end{Bmatrix}$

Using normalization

$$[d] = \begin{bmatrix} 0.4472 & -0.8944 \\ 0.8944 & 0.4472 \end{bmatrix}$$

$$[S] = [M]^{-\frac{1}{2}}\{d\} = \begin{bmatrix} 0.4472 & -0.8944 \\ 0.4472 & 0.2236 \end{bmatrix}$$

$$[S]^{-1} = \begin{bmatrix} 0.4472 & 1.7889 \\ -0.8944 & 0.8944 \end{bmatrix} = 2[d]^T[I]$$

$$\{r(0)\} = [S]^{-\frac{1}{2}}\{x_0\} = [\downarrow]\begin{Bmatrix} 0.01 \\ 0 \end{Bmatrix} = \begin{Bmatrix} 0.004472 \\ -0.008944 \end{Bmatrix}$$

$$\{\dot{r}(0)\} = \{0\}$$

$$\begin{bmatrix} 1 & 0 \\ 0 & 1 \end{bmatrix}\begin{Bmatrix} \ddot{r}_1 \\ \ddot{r}_2 \end{Bmatrix} + \begin{bmatrix} 0 & 0 \\ 0 & 5 \end{bmatrix}\begin{Bmatrix} r_1 \\ r_2 \end{Bmatrix} = 0$$

Zero Eigen Value

$$\ddot{r}_1(t) = 0$$

$$\dot{r}_1(t) = b$$

$$r_1(t) = a + bt$$

$$r_1(0) = a = 0.004772$$

$$\dot{r}_1(0) = b = 0$$

$$r_1(t) = 0.004772$$

$$r_2(t) = -0.0089\cos\left(\sqrt{5}t\right)$$

$$r_2(t) = \frac{\sqrt{r_{20}^2 \omega_{n_2}^2 + \dot{r}_2^2}}{\omega_{n_2}} \cdot Sin\left(\omega_{n_2}t + \tan^{-1}\frac{\omega_{n_2}r_{20}}{\dot{r}_{20}}\right)$$

$$\{x(t)\} = [S]\{r(t)\}$$

Hence, $= [S]\begin{Bmatrix} 0.004772 \\ -0.0089\,Cos\,\sqrt{5}t \end{Bmatrix}$

$$= \begin{bmatrix} 2.012 + 7.960\,Cos\,\sqrt{5}t \\ 2.012 - 1.990\,Cos\,\sqrt{5}t \end{bmatrix} \times 10^{-3}\,m$$

Both the bodies move by a distance 2.012mm(rigid body). The other part of motion is vibrating motion.

$$m\ddot{x} + Kx = 0 \quad x(0) = x_0 \quad \dot{x}(0) = x_0$$

Response

$$x(t) = \frac{\dot{x}(0)}{\omega_n}\sin\omega_n t + x(0)\,Cos\,\omega_n t$$

$$x(t) = \left[\left(\frac{\dot{x}(0)}{\omega_n}\right)^2 + (x(0))^2\right]^{\frac{1}{2}} Sin(\omega_n t + \alpha)$$

$$\alpha = \tan^{-1}\frac{\omega_n x(0)}{\dot{x}(0)}$$

Note : Define

$$A\,Sin\,\alpha =$$

$$\frac{\dot{x}(0)}{\omega_n} = A\,Cos\,\alpha \qquad and \qquad x(0) = A\,Sin\,\alpha$$

$$x(t) = A\left(Cos\,\alpha\,Sin\,\omega_n t + Cos\,\omega_n t\,Sin\,\alpha\right)$$

$$= A\,Sin\left(\omega_n t + \alpha\right)$$

From above equation

$$A^2 \left(Sin^2 \alpha + Cos^2 \alpha \right) = \left(\frac{\dot{x}(0)}{\omega_n} \right)^2 + \left(x(0) \right)^2$$

$$A = \left[\left(\frac{\dot{x}(0)}{\omega_n} \right)^2 + \left(x(0) \right)^2 \right]$$

$$\tan \alpha = \frac{Sin \alpha}{Cos \alpha} = \frac{\omega_n x(0)}{\dot{x}(0)}$$

$$\alpha = \tan^{-1} \left[\frac{\omega_n x(0)}{\dot{x}(0)} \right]$$

With viscous damping (under damped)

$$\ddot{r}_i(t) + 2\xi_i \, \omega_{n_i} \dot{r}_i(t) + \omega_{n_i}^2 r_i(t) = i = 1, 2, \ldots\ldots\ldots n$$

extra term entered for viscous damping

Damped Single Degree of Freedom System

$$m\ddot{x} + c\dot{x} + kx = 0 \qquad x(0) = x_0, \dot{x}(0) = v_0$$

$$\Rightarrow \ddot{x} + 2\xi\omega_n\dot{x} + \omega_n^2 x = 0 \qquad 0 < \xi < 1$$

Solution of this equn.

$$x(t) = A e^{-\xi\omega_n t} \sin(\omega_d t + \alpha_1)$$

$$\uparrow$$

damped freq.

$$\omega_d = \omega_n \sqrt{1 - \xi^2} \; ; \omega_n = \sqrt{\frac{k}{m}} \; ; \xi = \frac{c}{2m\omega_n}$$

$$A = \left[\frac{\left(\dot{x}_0 + \xi\omega_n x_0 \right)^2 + \left(x_0 \, \omega_d \right)^2}{\omega_d^2} \right]^{\frac{1}{2}}$$

$$\alpha_1 = \tan^{-1} \frac{x_0 \omega_d}{\dot{x}_0 + \xi\omega_n x_0}$$

Now, ω_{n_1} – natural frequency

Solution

$$r_i(t) = A_i e^{-\zeta_i \omega_{n_i} t} \sin\left(\omega_{1_i} t + \phi_i\right)$$

$$i = 1, 2, \ldots\ldots\ldots\ldots n$$

A_i and ϕ_i are determined from initial condition

$$\omega_{d,i} = \omega_{n_i} \sqrt{1 - \xi_i^2}$$

$$A_i = \left[\frac{\left(\dot{r}_{i0} + \xi_i \omega_{ni} r_{i0}\right)^2 + \left(r_{i0} \omega_{ni}\right)^2}{\omega_{1i}^2}\right]^{\frac{1}{2}}$$

$$\varphi_i = \tan^{-1} \frac{r_{i0} \omega_{1i}}{\dot{r}_{i0} + \xi_i \omega_{ni} r_{i0}}$$

Damped 2 DOF System

$$[M] = \begin{bmatrix} 9 & 0 \\ 0 & 1 \end{bmatrix} \qquad K = \begin{bmatrix} 27 & -3 \\ -3 & 3 \end{bmatrix} \qquad \xi_1 = 0.05, \xi = 0.10\} \text{ Modal Damping}$$

Initial Condition

$$\{x(0)\} = \begin{Bmatrix} 1 \\ 0 \end{Bmatrix}; \{\dot{x}(0)\} = \begin{Bmatrix} 0 \\ 0 \end{Bmatrix}$$

$$J_i = 0.05 \text{ and } \xi = 0.1$$

$$[M]^{-1/2} = \begin{bmatrix} \dfrac{1}{3} & 0 \\ 0 & 1 \end{bmatrix} \qquad \bar{K} = \begin{bmatrix} 3 & -1 \\ -1 & 3 \end{bmatrix}$$

$$[d] = \frac{1}{\sqrt{2}} \begin{bmatrix} 1 & 1 \\ 1 & -1 \end{bmatrix} [\Omega] = diag(2, 4)$$

$$\omega_{n1} = \sqrt{2} \qquad \omega_{n2} = 2$$

$$[S] = [M]^{-\frac{1}{2}} [d] = \frac{1}{\sqrt{2}} \begin{bmatrix} \dfrac{1}{3} & \dfrac{1}{3} \\ 1 & -1 \end{bmatrix}$$

$$[S]^{-1} = [d]^T [M]^{\frac{1}{2}} = \frac{1}{\sqrt{2}} \begin{bmatrix} 3 & 1 \\ 3 & -1 \end{bmatrix}$$

$$[S].[S]^{-1} = [I] \rightarrow [S]^T [M][S] = [I]$$

$$\{\dot{r}(0)\} = [S]^{-1}\{x_0\} = \begin{Bmatrix} 3/\sqrt{2} \\ 3/\sqrt{2} \end{Bmatrix}$$

$$\{\dot{r}(0)\} = [S]^{-1}\{\dot{x}_0\} = [S]^{-1} \begin{Bmatrix} 0 \\ 0 \end{Bmatrix} = \begin{Bmatrix} 0 \\ 0 \end{Bmatrix}$$

$$\begin{Bmatrix} r_{10} \\ r_{20} \end{Bmatrix} = \begin{Bmatrix} 3/\sqrt{2} \\ 3/\sqrt{2} \end{Bmatrix} \text{ and } \begin{Bmatrix} \dot{r}_{10} \\ \dot{r}_{20} \end{Bmatrix} = \begin{Bmatrix} 0 \\ 0 \end{Bmatrix}$$

$$\omega_{11} = 1.4124 \qquad \omega_{12} = 1.9900$$

$$A_1 = 2.1240 \qquad A_2 = 2.1320$$

$$\phi_1 = 1.52\,rad \qquad \phi_2 = 1.47\,rad$$

$$\{x(t)\} = [S]\{r(t)\} = \frac{1}{\sqrt{2}} \begin{bmatrix} 1 & 1 \\ 3 & 3 \\ 1 & -1 \end{bmatrix} \begin{Bmatrix} 2.124e^{-0.0706t}\sin(1.4124t+1.52) \\ 2.1320e^{-0.2t}\sin(1.9900t+1.4) \end{Bmatrix}$$

$$\{x(t)\} = \begin{bmatrix} 0.5006e^{-0.0706t}\sin(1.4124t+1.52)+0.525e^{-0.2t}\sin(1.9900t+1.4t) \\ 1.5019e^{-0.0706t}\sin(1.4124t1.52)-1.5076e^{-0.2t}\sin(1.9900t+1.4t) \end{bmatrix}$$

Response time curve

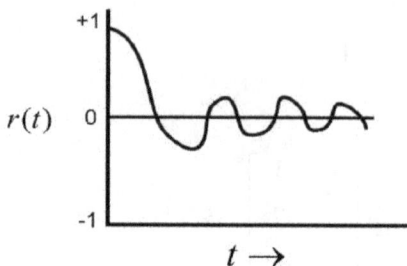

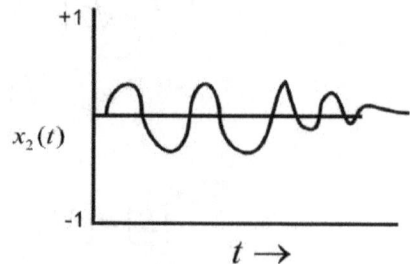

Forced Response – Modal Approach

$$[M]\{\ddot{x}\}+[C]\{\dot{x}\}+[K]\{x\}=[B]\{F(t)\}$$

$$[B]=\begin{bmatrix} 1 & 0 & 0 & ......0 \\ 0 & 1 & 0 & ......0 \\ 0 & 0 & 1 & ......0 \\ ... & ... & ... & ......1 \end{bmatrix}_{n\times n}$$

Assume

$$\{x(t)\}=[M]^{-1/2}\{q(t)\}$$

$$[I]\{\ddot{q}(t)\}+[\overline{C}]\{\dot{q}(t)\}+[\overline{K}]\{q(t)\}=[M]^{-1/2}[B]\{F(t)\}$$

$$[\overline{C}]=[M]^{-1/2}[C][M]^{-1/2}$$

$$[\overline{K}]=[M]^{-1/2}[K][M]^{-1/2}$$

Let $\{q(t)\}=[d]\{r(t)\}$

Substituting above relation and premultiply by $[d]^T$ we can write

$$\{\ddot{r}(t)\}+diag[2\xi_i\,\omega_{ni}]\{\dot{r}(t)\}+[\Omega]\{r(t)\}=[d]^T[M]^{-1/2}[B]\{F(t)\}$$

or, we can write

$$\ddot{r}_i(t)+2\xi_i\,\omega_{ni}\,\dot{r}_i(t)+\omega_{ni}^2 r_i(t)=f_i(t)$$

Assuming zero initial condition

$$x(t)=\frac{1}{m\omega_d}e^{-\xi\omega_n t}\int_0^t F(\tau)e^{\xi\omega_n\tau}\sin\omega_d(t-\tau)d\tau$$

Where $\omega_n=\sqrt{\dfrac{K}{m}}$ $\omega_d=\omega_n\sqrt{1-\xi^2}\,\xi=\dfrac{C}{2m\omega_n}$

For non-zero initial conditions

$$x(t)=Ae^{-\xi\omega_n t}\sin(\omega_d t+\phi)+\frac{1}{m\omega_d}e^{-\xi\omega_n t}\int_0^t F(\tau)e^{+\xi\omega_n\tau}\sin\omega_d(t-\tau)$$

Similarly

We can get modal solution

$$r_i(t) = A_i e^{-\xi_i \omega_{ni} t} \sin(\omega d_i t + \phi) + \frac{1}{\omega_{d_i}} e^{-\xi_i \omega_{ni} t} \int_0^t F_i(\tau) e^{\zeta_i \omega_{ni} \tau} \sin \omega_{d_i}(t - \tau) d\tau$$

$$\omega_{d_i} = \omega_{n_i} \sqrt{1 - \xi_i^2}$$

Ex

$$F_2(t) = 3 \cos 2t$$

$$m_1 = 9kg \ k_1 = 24N/m$$

$$m_2 = 1kg \ k_2 = 3N/m$$

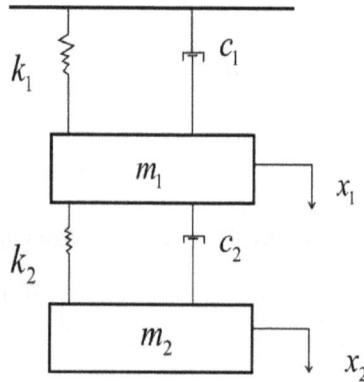

$$C_1 = 2.4N - S/m; \qquad C_2 = 0.3 \, N - S/m$$

The equilibrium equation in matrix form

$$\begin{bmatrix} 9 & 0 \\ 0 & 1 \end{bmatrix} \{\ddot{x}\} + \begin{bmatrix} 2.7 & -0.3 \\ -0.3 & 0.3 \end{bmatrix} \{\dot{x}\} + \begin{bmatrix} 27 & -3 \\ -3 & 3 \end{bmatrix} \{x\} = \begin{bmatrix} 0 & 0 \\ 0 & 1 \end{bmatrix} \begin{Bmatrix} 0 \\ F_2(t) \end{Bmatrix}$$

$$[M]^{\frac{1}{2}} = \begin{bmatrix} 3 & 0 \\ 0 & 1 \end{bmatrix} [M]^{-\frac{1}{2}} = \begin{bmatrix} 1/3 & 0 \\ 0 & 1 \end{bmatrix}$$

$$\bar{C} = \begin{bmatrix} 0.3 & -0.1 \\ -0.1 & 0.3 \end{bmatrix} \bar{K} = \begin{bmatrix} 3 & -1 \\ -1 & 3 \end{bmatrix}$$

$$\lambda_1 = 2, \lambda_2 = 4 \rightarrow \omega_{n_1} = \sqrt{2} \text{ and } \omega_{n_2} = 2$$

$$[d] = 0.7071 \begin{bmatrix} 1 & -1 \\ 1 & 1 \end{bmatrix}$$

$$d^T \bar{C} d = (0.7071)^2 \begin{bmatrix} 1 & 1 \\ -1 & 1 \end{bmatrix} \begin{bmatrix} 0.3 & -0.1 \\ -0.1 & 0.3 \end{bmatrix} \begin{bmatrix} 1 & -1 \\ 1 & 1 \end{bmatrix}$$

$$= (0.7071)^2 \begin{bmatrix} 1 & 1 \\ -1 & 1 \end{bmatrix} \begin{bmatrix} 0.2 & -0.4 \\ 0.2 & 0.4 \end{bmatrix}$$

$$= (0.7071)^2 \begin{bmatrix} 0.4 & 0 \\ 0 & 0.8 \end{bmatrix}$$

$$= \begin{bmatrix} 0.2 & 0 \\ 0 & 0.4 \end{bmatrix}$$

$$[f(t)] = [d]^T [M]^{-\frac{1}{2}} [B]\{F(t)\} \quad \bigg| \quad 2\varsigma\omega_n = 0.2$$
$$= \begin{bmatrix} 0.2357 & 0.7071 \\ -0.2357 & 0.7071 \end{bmatrix} \begin{Bmatrix} 0 \\ F_2(t) \end{Bmatrix} \bigg| \varsigma_1 = \frac{0.2}{2\omega_{n1}} = 0.0707$$
$$= 0.7071 \begin{Bmatrix} F_2(t) \\ F_2(t) \end{Bmatrix} \quad \bigg| \quad \varsigma_2 = \frac{0.4}{2\omega_2} = 0.10$$

$$\ddot{r}_1 + 0.2\dot{r}_1 + 2r_1 = 0.7071(3)\cos 2t = 2.1713\cos 2t$$

$$\ddot{r}_2 + 0.4\dot{r}_2 + 4r_2 = 2.1713\cos 2t$$

$$r_1(t) = ?? \rightarrow x_1(t) =$$
$$r_2(t) = ?? \rightarrow x_2(t) =$$

References

- Tan, Delin; Chen, Zheng (2012), "On A General Formula of Fourth Order Runge-Kutta Method" (PDF), Journal of Mathematical Science & Mathematics Education, 7.2: 1–10

- Ascher, Uri M.; Petzold, Linda R. (1998), Computer Methods for Ordinary Differential Equations and Differential-Algebraic Equations, Philadelphia: Society for Industrial and Applied Mathematics, ISBN 978-0-89871-412-8

- Stoer, Josef; Bulirsch, Roland (2002), Introduction to Numerical Analysis (3rd ed.), Berlin, New York: Springer-Verlag, ISBN 978-0-387-95452-3

- Butcher, John C. (May 1963), Coefficients for the study of Runge-Kutta integration processes, 3 (2), pp. 185–201, doi:10.1017/S1446788700027932

- "Le destin douloureux de Walther Ritz (1878-1909)", (Jean-Claude Pont, editor), Cahiers de Vallesia, 24, (2012), ISBN 978-2-9700636-5-0

- Ramin Shamshiri and Wan Ishak Wan Ismail, 2014. Implementation of Galerkin's Method and Modal Analysis for Unforced Vibration Response of a Tractor Suspension Model. Research Journal of Applied Sciences, Engineering and Technology. 7(1): 49-55

- S. Brenner, R. L. Scott, The Mathematical Theory of Finite Element Methods, 2nd edition, Springer, 2005, ISBN 0-387-95451-1

- Dahlquist, Germund (1963), "A special stability problem for linear multistep methods", BIT, 3: 27–43, ISSN 0006-3835, doi:10.1007/BF01963532

Permissions

All chapters in this book are published with permission under the Creative Commons Attribution Share Alike License or equivalent. Every chapter published in this book has been scrutinized by our experts. Their significance has been extensively debated. The topics covered herein carry significant information for a comprehensive understanding. They may even be implemented as practical applications or may be referred to as a beginning point for further studies.

We would like to thank the editorial team for lending their expertise to make the book truly unique. They have played a crucial role in the development of this book. Without their invaluable contributions this book wouldn't have been possible. They have made vital efforts to compile up to date information on the varied aspects of this subject to make this book a valuable addition to the collection of many professionals and students.

This book was conceptualized with the vision of imparting up-to-date and integrated information in this field. To ensure the same, a matchless editorial board was set up. Every individual on the board went through rigorous rounds of assessment to prove their worth. After which they invested a large part of their time researching and compiling the most relevant data for our readers.

The editorial board has been involved in producing this book since its inception. They have spent rigorous hours researching and exploring the diverse topics which have resulted in the successful publishing of this book. They have passed on their knowledge of decades through this book. To expedite this challenging task, the publisher supported the team at every step. A small team of assistant editors was also appointed to further simplify the editing procedure and attain best results for the readers.

Apart from the editorial board, the designing team has also invested a significant amount of their time in understanding the subject and creating the most relevant covers. They scrutinized every image to scout for the most suitable representation of the subject and create an appropriate cover for the book.

The publishing team has been an ardent support to the editorial, designing and production team. Their endless efforts to recruit the best for this project, has resulted in the accomplishment of this book. They are a veteran in the field of academics and their pool of knowledge is as vast as their experience in printing. Their expertise and guidance has proved useful at every step. Their uncompromising quality standards have made this book an exceptional effort. Their encouragement from time to time has been an inspiration for everyone.

The publisher and the editorial board hope that this book will prove to be a valuable piece of knowledge for students, practitioners and scholars across the globe.

Index

www.ingramcontent.com/pod-product-compliance
Lightning Source LLC
Chambersburg PA
CBHW061956190326
41458CB00009B/2884